D1739838

Please return to:

WITHDRAWN

PHARMA
R & D LIBRARY

PHARMACEUTICAL SOLID DISPERSION TECHNOLOGY

HOW TO ORDER THIS BOOK

BY PHONE: 800-233-9936 or 717-291-5609, 8AM–5PM Eastern Time

BY FAX: 717-295-4538

BY MAIL: Order Department
Technomic Publishing Company, Inc.
851 New Holland Avenue, Box 3535
Lancaster, PA 17604, U.S.A.

BY CREDIT CARD: American Express, VISA, MasterCard

BY WWW SITE: http://www.techpub.com

PERMISSION TO PHOTOCOPY—POLICY STATEMENT

Authorization to photocopy items for internal or personal use, or the internal or personal use
of specific clients, is granted by Technomic Publishing Co., Inc. provided that the base fee of
US $5.00 per copy, plus US $.25 per page is paid directly to Copyright Clearance Center,
222 Rosewood Drive, Danvers, MA 01923, USA. For those organizations that have been
granted a photocopy license by CCC, a separate system of payment has been arranged.
The fee code for users of the Transactional Reporting Service is 1-58716/01 $5.00 + $.25.

Pharmaceutical Solid Dispersion Technology

EDITED BY

Muhammad J. Habib, Ph.D.

School of Pharmacy
Howard University
Washington, D.C.

Please return to:

PHARMA
R & D LIBRARY

TECHNOMIC
PUBLISHING CO., INC.
LANCASTER • BASEL

Pharmaceutical Solid Dispersion Technology
aTECHNOMIC®publication

Technomic Publishing Company, Inc.
851 New Holland Avenue, Box 3535
Lancaster, Pennsylvania 17604 U.S.A.

Copyright © 2001 by Technomic Publishing Company, Inc.
All rights reserved

No part of this publication may be reproduced, stored in a
retrieval system, or transmitted, in any form or by any means,
electronic, mechanical, photocopying, recording, or otherwise,
without the prior written permission of the publisher.

Printed in the United States of America
10 9 8 7 6 5 4 3 2 1

Main entry under title:
   Pharmaceutical Solid Dispersion Technology

A Technomic Publishing Company book
Bibliography: p.
Includes index p. 95

Library of Congress Catalog Card No. 00-107623
ISBN No. 1-56676-813-6

To my Mom and Dad

Table of Contents

Preface

SOLID dispersion technology is the science of dispersing one or more active ingredients in an inert matrix in the solid state in order to achieve increased dissolution rate, sustained release of drugs, altered solid-state properties, enhanced release of drugs from ointment and suppository bases, and improved solubility and stability. Today, this technology encompasses not only the traditional coprecipitates but also certain nanoparticles, microcapsules, microspheres and other dispersions of drugs in polymers prepared by using any one of the above processes. It has been about 25 years since the first report on the concept and advantages of solid dispersion formulations was published by Sekiguchi and Obi. They demonstrated that a eutectic mixture of sulfathiazole and urea had a much higher dissolution rate leading to better absorption following oral administration as compared to sulfathiazole administered alone. This presented a novel method for reducing the particle size of poorly soluble drugs. Subsequently, numerous research articles dealing with solid dispersions and a few review articles have been published, but the scientific community still has not seen a complete book on the subject, dealing with all the aspects of solid dispersions in a concise manner. This book is an easy reference tool, useful for solving numerous problems of drug formulations and bioavailability.

The overall aims and objectives of the book are to discuss solid dispersion from a practical point of view. Necessary theoretical concepts have been provided to assist in comprehending the subject matter. This book is designed for individuals in industry who are solving various formulation problems day to day. This book should also be helpful to individuals in academia and in research by providing theoretical and up-to-date developments in research related to this field of endeavor.

Chapter 1 pertains to the history and background of solid dispersions. In Chapter 2, the fundamentals and the theoretical aspects are discussed. The mathematical approach and various dissolution laws are also discussed here. The physicochemical characteristics of the drugs and the solvent systems are introduced, and the various theories and mechanisms of solid dispersions elaborated. These chapters will be helpful to those individuals who are new to the field and would like to understand more about the properties of solid dispersion technology. The next few chapters are devoted mainly to the various techniques and methods of solid dispersions, namely, polymers, cyclodextrins, and phospholipids. The intention is to incorporate in this book all the latest developments in the field.

Acknowledgements

ONE person has been very helpful in preparing this book. I thank him sincerely for extending my own understanding. Dr. James Rogers introduced me to many concepts of solid dispersion and inspired me to undertake this venture. I also thank all the contributors to this book, who, although hard-pressed, agreed to submit their manuscripts on time. Finally, I thank my wife Oliza Khanam for her love, encouragement, and support over these two years of preparation.

CHAPTER 1

Historical Background of Solid Dispersions

MUHAMMAD J. HABIB[1]

W HEN a drug is administered perorally in a solid dosage form such as tablet, capsule, or suspension it must be released from the dosage form and dissolved in the gastrointestinal fluids before it can be absorbed. The bioavailability of many poorly water-soluble drugs is limited by their dissolution rates, which are in turn controlled by the surface area that they present for dissolution. The effect of the particle size of drugs on their dissolution rates and biological availability was reviewed comprehensively by Fincher [1]. Two consecutive transport processes can be identified to describe the oral absorption of drugs from solid dosage forms: (1) dissolution of the drug *in vivo* to produce a solution; and, (2) transport of the dissolved drug across the gastrointestinal membrane. Each process can be characterized by a rate constant. If the rate of dissolution of the drug is significantly slower than the rate of absorption, the dissolution of the drug becomes the rate-limiting step in the absorption process, and the particle size of the drug is of great importance in the transport from the gastrointestinal (GI) tract to the site of action. Most drugs are passively absorbed and their rates of absorption are dependent upon the concentration gradients in each case; by increasing the dissolution rate in the GI tract, the absorption rate is increased so long as the dissolution rate is still the limiting step [2]. This commonly occurs for drugs with limited water solubility.

[1]School of Pharmacy, Howard University, Washington, D.C., U.S.A.

For example, the therapeutic dose of griseofulvin was reduced to 50% by micronization [3], and it also produced a more constant and reliable blood level. The stability characteristics of the drug may be altered considerably by reduction of particle size. The resultant increase in surface area places more of the drug molecules in a vulnerable position for rapid degradation by the GI fluid [4]. The degree of agitation in the GI contents is related to particle-size effects, and this factor is subject to considerable variability within the individual, depending on the physiological state during the time of administration and absorption. The agitation intensity would affect the size of the "saturation boundary layer" on each of the drug particles and also the "effective diffusion rate" of the drug to the sites of the absorption [5].

In parenteral therapy, the particle size of suspended particles injected intramuscularly or subcutaneously is an important factor. The absorption of the drugs from particles appears to increase with an increase in the specific surface area; however, in certain vehicles where a hydrophobic protective layer is involved, the absorption may be delayed when using very small particles. This may be related to the fact that fine particles tend to cake or form gels in certain vehicles. The viscosity of suspensions increases with a decrease in particle size, and this factor may explain some of the delayed absorption from using very small particles [6,7].

In semisolid systems (ointments and suppositories), the particle size of the insoluble fraction of the drug is an important factor in obtaining the desired results. In these cases an improved dissolution rate leads to greater availability of the drug for absorption in the case of systemic effects, and for higher concentration at the area of application for a better local action [6,7].

Inhalation therapy has been improved considerably by controlling the particle size of aerosol-generated droplets. The size of the droplet governs the deposition area in the respiratory tract [6,7]. The particle size of practically insoluble drugs in dusting powders, insufflates, and aerosols was predicted to be an important consideration in their dissolution and subsequent local action [6].

Particle-size reduction is usually achieved by: (1) conventional trituration and grinding; (2) ball milling; (3) fluid energy micronization; (4) controlled precipitation by change of solvents or temperature, application of ultrasonic waves [8–10], and spray drying [11]; (5) administration of liquid solutions from which, upon dilution with gastric fluids, the dissolved drug may precipitate in very fine particles [12]; and (6) administration of water-soluble salts of poorly soluble compounds from which the parent neutral forms may precipitate in ultrafine form in GI fluids. Although the reduction of particle size can be easily and directly accomplished by the first four methods (1–4), the resultant fine particles may not produce expected faster dissolution and absorption. This primarily results from the possible aggregation and agglomeration of the fine particles due to their increased surface energy and the subsequent stronger van der waals' attraction between nonpolar molecules. This was demonstrated by Lin et al. [13], who showed that the *in vitro* dissolution rates of micronized griseofulvin and glutethimide were slower than those of their coarser particles. However, the opposite finding for griseofulvin was reported by Chiou and Riegelman [3]. Another inherent disadvantage of these pure, fine powders of poorly soluble drugs is their poor wettability in water. The wetting of powders is the first step for them to dissolve and sometimes disperse in fluids. Furthermore, drugs with plastic properties are difficult to subdivide by methods 1–3. They have more tendency to stick together, even if fine powders can be produced by controlled precipitation.

Theoretically, the solvent method (3) seems to be an ideal approach to achieving particle-size reduction. However, it is not frequently employed in the commercial market due to such reasons as selection of a nontoxic solvent, limitation to drugs with low dose, and the high costs of production. The water-soluble salts of many poorly soluble acidic or basic drugs have been widely used clinically as solid dosage forms. Indeed, they have been shown frequently to produce better absorption than their parent forms. It has been demonstrated that the potassium or sodium salts may react with atmospheric carbon dioxide and water to precipitate out poorly soluble parent compounds. This occurs specially on the outer layer of a dosage form and thereby retards rates of dissolution and absorption. This precipitate effect is believed to be responsible for the slower *in vitro* dissolution rates and the lower novobiocin plasma levels in dogs following oral administration of its soluble sodium salt rather than the less soluble amorphous form of the parent compound [14]. The reported failure of the clinical response from three commercial capsule dosage forms containing sodium diphenylhydantoin may be caused for the same reason [15]. In addition, the alkalinity of some salts may cause epigastric distress following administration.

In 1961, a unique approach of solid dispersion to reduce the particle size and increase rates of dissolution and absorption was first demonstrated by Sekiguchi and Obi [2]. They proposed the formation of a eutectic mixture of a poorly soluble drug such as sulfathiazole with a physiologically inert, freely water-soluble carrier such as urea. The eutectic mixture was prepared by melting the physical mixture of the drug and the carrier, followed by a rapid solidification process. Upon exposure to aqueous fluids, the active drug was expected to be released into the fluids as fine, dispersed particles because of the fine dispersion of the drug in the solid eutectic mixture and the rapid dissolution of the soluble matrix. Levy [12] and Kanig [16] subsequently noted the possibility of using a solid solution approach in which a drug is dispersed molecularly in a soluble carrier. In a series of reports in 1965–1966, Goldberg et al. [17–20] presented a detailed experimental and theoretical discussion of the advantages of solid solution over the eutectic mixture.

In 1965, Tachibana and Nakamura [21] reported a novel method of preparing aqueous colloidal dispersions of β-carotene by using water-soluble polymers such as polyvinylpyrrolidone. They dissolved the drug and the polymer carrier in a common solvent and then evaporated the solvent completely. A colloidal dispersion was obtained when the coprecipitate was exposed to water. In 1966, Mayersohn and Gibaldi [5] demonstrated that the dissolution rate of griseofulvin could be markedly enhanced when dispersed in polyvinylpyrrolidone by the same solvent method. The mechanisms of increased dissolution rates of drugs, solid dispersed in polyvinylpyrrolidone carriers, were thoroughly discussed by Simonelli et al. [4]. Chiou and Riegelman [3] recently advocated the application of glass solution to increase dissolution rates. The significance of the solid dispersion technique was strengthened by the demonstration of Chiou and Riegelman [22] of the fast and almost complete absorption of the insoluble griseofulvin in man and dogs while the commercial micronized griseofulvin was incompletely absorbed (30–60%). They used polyethylene glycol 6000 as a dispersion carrier. The main advantages of using water-soluble polymers as carriers are their nontoxicity and general applicability to most drugs.

It is believed that this field of pharmaceutical technique and principles will play an important role in increasing dissolution, absorption, and therapeutic efficacy of drugs in future dosage forms. Therefore, a thorough understanding of its fast-release principles, method of preparation, selection of suitable carriers, determination of physical properties, limitations, and disadvantages will be essential in the practical and effective application of this approach. The main purpose of this book is to review critically the pharmaceutical literature pertinent to this area. In addition to absorption enhancement, the solid dispersion technique has numerous other pharmaceutical applications that will also be explored. For example, such a technique can be used to obtain a homogeneous distribution of a small amount of drugs at solid state, to stabilize unstable drugs, to dispense liquid or gaseous compounds, to formulate a fast-release priming dose in a sustained-release dosage form, and to formulate sustained-release or prolonged-release regimens of soluble drugs by using poorly soluble or insoluble carriers. It is hoped that this book will stimulate interest and research in these various areas.

REFERENCES

1 Fincher, J. H. 1968. "Particle size of drugs and its relationship to absorption and activity." *J. Pharm. Sci.,* 57(11):1825–35.

2 Sekiguchi, K. and Obi, N. 1961. "Studies on absorption of eutectic mixtures. I. A comparison of the behavior of eutectic mixture of sulfathiazole and that of ordinary sulfathiazole in man." *Chem. Pharm. Bull.,* 9:866–72.

3 Chiou, W. L. and Riegelman, S. 1969. "Preparation and dissolution characteristics of several fast-release solid dispersions of griseofulvin." *J. Pharm. Sci.,* 58(12):1505–10.

4 Simonelli, A. P., Mehta, S. C. and Higuchi, W. I. 1969. "Dissolution rates of high energy polyvinylpyrrolidone (PVP)-sulfathiazole coprecipitates." *J. Pharm. Sci.,* 58(5):538–49.

5 Mayersohn, M. and Gibaldi, M. 1966. "New method of solid-state dispersion for increasing dissolution rates." *J. Pharm Sci.,* 55(11):1323–4.

6 Lees, K. A. 1963. "Pharmaceutical aspects of fine particles and their evaluation." *Pharm. J.,* 191:289–91.

7 Dare, J. G. 1964. "Particle size in relation to formulation." *Australian J. Pharm.,* 45:S58–S65.

8 Scheikh, M. A., Price, J. C. and Gerraughty, R. J. 1966. "Effect of ultrasound on particle size suspensions of polyethylene spheres." *J. Pharm. Sci.,* 55:1048–50.

9 Hem, S. L., Skauen, D. M. and Beal, H. M. 1967. "Mechanism of crystallization of hydrocortisone by ultrasonic irradiation." *J. Pharm. Sci.,* 56(2):229–33.

10 Skauen, D. M. 1967. "Some pharmaceutical applications of ultrasonics." *J. Pharm Sci.,* 56(11):1373–85.

11 Kornblum, S. S. and Hirschorn, J. O. 1970. "Dissolution of poorly water-soluble drugs. I. Some physical parameters related to method of micronization and tablet manufacture of a quinazolinone compound." *J. Pharm. Sci.,* 59(5):606–9.

12 Levy, G. 1963. "Effect of particle size on dissolution and gastrointestinal absorption rates of pharmaceuticals." *Amer. J. Pharm.*, 135:78–92.

13 Lin, S. L., Menig, J. and Lachman, L. 1968. "Interdependence of physiological surfactant and drug particle size on the dissolution behavior of water insoluble drugs." *J. Pharm. Sci.*, 57(12):2143–8.

14 Mullins, J. D. and Macek, T. J. 1960. "Some pharmaceutical properties of novobiocin." *J. Amer. Pharm. Ass.*, 49:245–8.

15 Feinberg, M. 1969. "Drug standards in military procurement." *J. Am. Pharm. Ass.*, 9(3):113–6.

16 Kanig, J. L. 1964. "Properties of fused mannitol in compressed tablets." *J. Pharm. Sci.*, 53:188–92.

17 Goldberg, A. H., Gibaldi, M. and Kanig, J. L. 1965. "Increasing dissolution rates and gastrointestinal absorption of drugs via solid solutions and eutectic mixtures. I. Theoretical considerations and discussion of the literature." *J. Pharm. Sci.*, 54:1145–8.

18 Goldberg, A. H., Gibaldi, M. and Kanig, J. L. 1966. "Increasing dissolution rates and gastrointestinal absorption of drugs via solid solutions and eutectic mixtures. II. Experimental evaluation of a eutectic mixture: urea-acetaminophen system." *J. Pharm. Sci.*, 55:482–7.

19 Goldberg, A. H., Gibaldi, M. and Kanig, J. L. 1966. "Increasing dissolution rates and gastrointestinal absorption of drugs via solid solutions and eutectic mixtures. III. Experimental evaluation of griseofulvin-succinic acid solid solution." *J. Pharm. Sci.*, 55:487–92.

20 Goldberg, A. H. and Gibaldi, M. 1966. "Increasing dissolution rates and gastrointestinal absorption of drugs via solid solutions and eutectic mixtures. IV. Chloramphenicol—urea system." *J. Pharm. Sci.*, 55(6): 581–3.

21 Tachibana, T. and Nakamura, N. 1965. "A method of preparing an aqueous colloidal dispersion of organic materials by using water-soluble polymers: dispersion of β-carotene by poly (vinyl pyrrolidinone)." *Kolloid-Z.polymer,* 203: 130–3.

22 Chiou, W. I. and Riegelman, S. 1970. "Oral absorption of griseofulvin in dogs: increased absorption via solid dispersion in polyethylene glycol 6000." *J. Pharm. Sci.*, 59, 937–41.

CHAPTER 2

Fundamentals of Solid Dispersions

MUHAMMAD J. HABIB[1]
SURESH VENKATARAM[2]
M. DELWAR HUSSAIN[3]

THEORY OF DISSOLUTION

DISSOLUTION rate may be defined as the amount of drug substance that is dissolved per unit time under standardized conditions of liquid-solid interface, temperature, and solvent composition. Dissolution can be considered a specific type of heterogeneous reaction in which a mass transfer results as a net effect between the escape and deposition of solute molecules at a solid surface. The most common theory for dissolution, the film theory, also known as the diffusion layer model, accepts the assumption that dissolution belongs to a type of heterogeneous reaction where the rate is determined by the transport process. The following is the brief interpretation of this as well as some other important dissolution theories.

NOYES-WHITNEY AND NERNST-BRUNNER EQUATIONS

Noyes and Whitney in 1897 [1] stated that the rate at which a solid substance dissolves in its own solution is proportional to the difference between the concentration of that solution and the concentration of the saturated solution. Mathematically it can be expressed as,

[1]School of Pharmacy, Howard University, Washington, D.C., U.S.A.
[2]Patheon, Inc., Mississauga, Ontario, Canada.
[3]College of Pharmacy, University of Wyoming, Laramie, WY, U.S.A.

$$\frac{dc}{dt} = K(C_s - C_b) \qquad (1)$$

where dc is the dissolution rate, K is a proportionality constant, C_s is the solubility of the solute, and C_b is the concentration at any time, t.

In the integral form Equation (1) can be used to calculate the value of the proportionality constant if the solubility is known. Thus,

$$K = \frac{1}{t} \ln \frac{C_s}{(C_s - C_b)} \qquad (2)$$

The Noyes-Whitney equation can be explained as: (1) a thin layer of saturated solution is formed at the surface of the solid and the rate of dissolution is governed by the rate of diffusion from this layer to the bulk of the solution, and (2) there is negligible change in the surface area with time during dissolution.

Noyes and Whitney [2] and Brunner and Tolloczko [3] revised the equation assuming, that under well-defined conditions of temperature and agitation, the dissolution rate is proportional to the surface area, S, giving

$$\frac{dc}{dt} = K_1 S(C_s - C_b) \qquad (3)$$

where K_1 is called the intrinsic dissolution rate constant.

Applying Fick's law of diffusion, Nernst [4] and Brunner [5] developed a modified form of this equation:

$$\frac{dc}{dt} = \frac{DS}{hV}(C_s - C_b) \qquad (4)$$

where, D is the diffusion coefficient of the solute, h is the thickness of the diffusion layer, and V is the volume of the dissolution medium. This has been referred to as the *film theory* of Nernst Brunner, which applies to some situations but not others [6].

CUBE ROOT LAW

Hixon and Crowell [7] introduced the concept of changing surface area during dissolution and derived the "cube root law" given by

$$(W_o)^{1/3} - (W_t)^{1/3} = \left(\frac{\pi N \rho}{6}\right)^{1/3} \frac{2DC_s t}{h\rho} \tag{5}$$

where, W_o is the initial weight of solid, W_t is the weight of solid at time, t, N is the number of particles, and ρ is the density of the solid. This equation is based on a number of assumptions: (1) dissolution takes place normal to the surface of the dissolving solid particle, (2) no stagnation of liquid occurs in any region, (3) the same effect of agitation is observed on all areas of the solid surface, (4) solid particles remain intact during the dissolution process, and (5) the stagnant or diffusion layer thickness is independent of the particle diameter.

Pedersen derived a general equation for characterizing drug powders based on single-particle dissolution according to the cube root or the square root law [8]:

$$Q^{1/(m-1)} = (mWK_m)^{1/(m-1)} - (mW)^{1/(m-1)} - (K_m)^{m/(m-1)}t \tag{6}$$

where Q is the dissolution rate, K_m is called specific dissolution rate parameter, W is weight of the powder and t is the time.

For a monodisperse power having a log normal distribution, the above equation predicts a linear relationship between $\sqrt{Q}$ and t when the particles dissolve according to cube root model ($m = 3$) and between Q and t when the particles dissolve according to the square root law ($m = 2$).

CONVECTIVE-DIFFUSION MODELS

The Hixon-Crowell model does not incorporate a quantitative consideration of the hydrodynamic effects. In order to overcome this deficiency, a model for a moving sphere was introduced by Nielsen [9], described by the equation

$$-\frac{da}{dt} = \frac{DC_s F}{a\rho} \tag{7}$$

where, a is the particle radius, $F = (1 + A)^{0.285}$ and

$$A = \frac{2a^3 g(\rho - \rho_0)}{9D\eta_0} \tag{8}$$

ρ_o is the density of the medium, η_o is the viscosity of the medium, and g is the acceleration due to gravity. This model takes into consideration a combined diffusion, convection-controlled dissolution rate of a particle and is based on the following assumptions: (1) the dissolution rate is diffusion-rate controlled, but convection also contributes to the transport process, (2) the effective diffusion layer thickness is the same for all particles of the same size and is equal to or greater than the particle radius, (3) particles are dissolving under sink conditions, and (4) the particles are spherical.

As the particle radius approaches zero, the quantity A approaches zero, the function F in Equation (7) approaches unity, and the equation reduces to that of a pure diffusional model, i.e., the Higuchi-Hiestand model [10].

$$-\frac{da}{dt} = \frac{DC_s}{a\rho} \tag{9}$$

Nelson and Shah [11] developed a mathematical model based on convective diffusion to describe the dissolution from the surface of a compressed compact. They evaluated the expression for two shapes—circular and rectangular—given by Equation (10) and (11), respectively

$$Q = 2.157D^{2/3}C_s\alpha^{1/3}r^{5/3} \tag{10}$$

$$Q = 0.808D^{2/3}C_s\alpha^{1/3}bL^{2/3} \tag{11}$$

where Q is the dissolution rate, b is the width of the tablet perpendicular to flow, L is the length of the tablet parallel to flow, α is the rate of shear in the boundary layer, and r is the radius of the tablet.

DANKWERTS MODEL

Dankwerts [12] proposed a model in which macroscopic packets reach the solid-liquid interface by eddy diffusion. He discarded the possibility of the existence of a stagnant layer. It was assumed that the surface is continually replaced with fresh liquid.

Mathematically this model is described by:

$$Q = S^{1/2}D^{1/2}A(C_s - C_b) \tag{12}$$

where, S is the rate of surface renewal and A is the area of the solid. Dankwerts originally applied this model to the dissolution of a gas in a liquid. However, it has been used by others to describe the dissolution of solids in multiparticulate systems [13].

DISSOLUTION KINETICS USING WEIBULL PLOT

Weibull [14] suggested a general function that is applicable to a number of common types of dissolution curve. The Weibull equation,

$$m = 1 - \exp[-(t - T_i)^b(1/a)] \tag{13}$$

where m is the accumulated fraction of the material in solution at time t, a is the scale parameter that defines the time scale of the process, T_i is the location parameter that represents the time lag before the onset of dissolution, and b is the shape parameter that characterizes the curve as being curved upwards ($b > 1$). The Weibull distribution functions can be arranged [15] in the form of a more useful equation as follows:

$$(1 - m) \exp[-(t - T_i)^b(1/a)]$$

$$-\ln (1 - m) = (t - T_i)^b(1/a)$$

$$\log [-\ln (1 - m)] = b \log (t - T_i) - \log a] \tag{14}$$

Using the linear relationship given by Equation (14), the shape parameter b and the scale parameter a can be obtained.

The scale parameter a is normally replaced by means of a more informative term, the dissolution time T_d, which is defined by

$$T_d = a^{1/b} \tag{15}$$

Equation (15) gives the time required to dissolve 63.2% of the drug. It can be also read from the graph directly as the time value corresponding to the ordinate value of zero. This condition is satisfied when $m = 0.63212$, i.e.,

$$\log [-\ln (1 - 0.63212)] = 0 \tag{16}$$

FIRST-ORDER KINETIC MODEL

The first-order rate equation to explain dissolution profiles is mathematically given by,

$$\log (W_\infty - W) = \log M - \frac{K}{2.303} (t - t_o) \qquad (17)$$

where, W_∞ is the amount of drug in solution at infinite time, W is the amount of drug in solution at time t, M is an integration constant, t_o is the lag time before the onset of dissolution, and K is the apparent first-order dissolution rate constant [16]. The equation may be slightly modified by using percent drug dissolved at time t as W, and using a value of 100 for W_∞. The time lag, t_o, for a powder formulation can be taken to be zero for all practical purposes. Thus Equation (17) becomes

$$\log (100 - \%\text{Dissolved}) = \log M - \frac{K t}{2.303} \qquad (18)$$

If the dissolution profile follows first-order kinetics, a straight line should result when log (100% dissolved) is plotted against t. The rate constant K is then obtained from the slope of the line.

SECOND-ORDER KINETIC MODEL

The second-order rate equation is given by,

$$\frac{W}{W_e (W_e - W)} = K_2 t \qquad (19)$$

where, W is the weight of the drug in solution at time t, W_e is the maximum amount of drug available for dissolution, and K_2 is the apparent second-order dissolution rate constant [17].

A plot of $W/W_e (W_e - W)$ against t yields a straight line if the dissolution follows second-order kinetics. The apparent second-order dissolution rate constant K_2 is given by slope of the line.

DISSOLUTION METHODS

In the past several years many attempts have been made to develop a reliable *in vitro* dissolution test that can thoroughly characterize the *in vivo* dissolution rate-controlled absorption of drugs. Criticisms in the early 1960s about the conventional tablet disintegration test as an indicator of drug release initiated an intense search for a better method. Researchers attempted to improve the disintegration test apparatus. However, by the mid-1960s it was suggested that the disintegration test be phased out in favor of a dissolution test. This led to the inclusion of dissolution standards for several monographs in the *USP XVIII* and *NF XIII*. Since then, the number of monographs required to meet the dissolution standards has increased to 400 in the *USP XXI*.

The rotating basket dissolution method was included as the first official compendial dissolution test (Method 1). The *USP XX* in 1980 included a modification of this method and also introduced (Method 2)—the paddle method—as an official compendial test. The disintegration test apparatus for dissolution (Method 3) has been deleted from the *USP XXI*.

A review of dissolution tests cited over 150 different apparatus designs [18]. The various methods differ in hydrodynamic properties, agitating intensities, and mechanical destructive forces to which the intact drug is subjected. One of the methods that has received acceptance by the scientific community is the Spin-Filter Stationery Basket apparatus developed by Shah and Nelson [19]. The apparatus has several advantages over the current USP methods. It consists of a dynamic *in situ* imcroporous, nonclogging rotating filter, which permits continuous and efficient filtration of the dissolution fluid. Furthermore, its large filter permits representative sampling of the bulk dissolution medium. The smooth cylindrical surface of the rotating filter assembly and its agitation of fluid over a long vertical axis provide uniform, mild, laminar, nonturbulent, and reproducible stirring at relatively high speeds. The spin filter apparatus was also found to give more reproducible results than the current USP method (the coefficient of variation being 3% and 42%, respectively) [19].

FACTORS AFFECTING DISSOLUTION RATE

In principle, variation of any of the parameters given in the dissolution equations, e.g., Equation (4), will cause a change in the rate of dissolution. However, in practice only the solubility and the surface area of the drug are actually controllable parameters. The diffusion layer thickness h is a function of the hydrodynamics of the system and, therefore, difficult to control. Also, the diffusion coefficient D cannot be made to vary except perhaps by slight changes in the viscosity of the dissolution medium. Hence, h and D are regarded as noncontrollable parameters in dosage form design.

SOLUBILITY OF THE DRUG IN DIFFUSION LAYER

Several methods of formulation can be used to alter the solubility of a drug in the theoretical diffusion layer.

SELECTION OF SALT FORM OF THE DRUG

The dissolution rate of the salt form of a drug would normally be expected to be greater than its nonsalt form, since the salt form is generally more soluble. However, the solubility of the salt depends on the counter-ion: the smaller the counter-ion, the more soluble is the salt. The sodium and potassium salts of para-amino-salicylic acid are approximately 1000 more times more soluble than unionized weak acid. Consequently, when these were administered to patients, higher blood levels of the ionized forms were obtained [20]. Similarly, tolbutamide sodium salt gave higher blood levels than the free acid [21]. In most of the salt forms studied, increased *in vitro* dissolution rate correlated well with the increased decline of blood sugar level. In some cases, the use of salt forms has decreased the absorption of the drug, e.g., aluminum acetyl salicylate, sodium warfarin, benzphetamine pamoate [22,23].

EFFECT OF PH

A majority of drugs are either weak acids or weak bases. This makes it convenient to increase the aqueous solubility of such drugs by adjusting the pH. A slight modification of the Nernst-Brunner equation shows that as the pH increases, the dissolution rate of a weak acid increases and of a weak base decreases. A classic example is the addition of buffering ingredients such as sodium bicarbonate to aspirin formulations in order to raise the pH of the microenvironment of the dissolving solid particle, which enhances its dissolution and thus its oral absorption [24].

CHANGE OF CRYSTAL FORM

Polymorphism describes the existence of a drug in two or more crystalline forms, each of which possesses a different space lattice arrangement but are otherwise chemically identical. Polymorphs exhibit different x-ray diffraction patterns, densities, melting points, stabilities, and solubilities [20]. A well-known example of the use of a polymorophic form to enhance bioavailability is that of chloramphenicol palmitate, which exists in three crystalline forms, A, B, and C, and one amorphous form [25]. The highest mean blood levels following oral dosing of the polymorphs to humans were obtained with Form B. Heleblian and McCrone [26] have reviewed the earlier literature with respect to pharmaceutical applications of polymorphism.

Solvate formation: solvates are crystalline forms of drugs combined with one or more molecules of solvent in the crystal lattice [27]. There appears to be no general rule concerning the dissolution of solvates formed from nonaqueous solvents compared to the nonsolvated form. Generally, the aqueous solubility and dissolution rate of the anhydrous form are greater than those of the hydrated forms [28]. For example, it has been shown that the anhydrous form of theophylline and cholesterol dissolve more rapidly than their hydrated forms [20]. Similarly, anhydrous ampicillin has a greater aqueous dissolution rate and yields greater area under the serum concentration-time curve than its corresponding trihydrate form [29]. Bates et al. [30] obtained an improved dissolution rate and oral absorption from the griseofulvin-chloroform solvate than from the nonsolvated micronized griseofulvin following oral administration to humans. A thorough review of drug solvate has been published by Haleblian [27].

Presence of crystal lattice imperfections has been shown to affect dissolution rate due to changes in the thermodynamic properties of crystals. Chiou and Kyle [31] have demonstrated the dissolution rate and the bioavailability of digoxin to be highly dependent on its crystals properties. Similar observations have been reported for adipic acid crystals by Chow et al. [32]. A positive correlation between the dissolution rate constant and the dislocation density has been reported for potassium perchlorate crystals [33]. A dimension-less term called *disruption index* has been proposed by York and Grant [34] to thermodynamically quantify the disorder imperfections in crystals induced by an impurity or additive.

COMPLEXATION

Interaction of a drug with a complexing agent may result in either an increase or a decrease in the solubility and dissolution rate of the drug. A drug may complex with both absorbable and nonabsorbable excipients in a dosage form. The dissolution rates of a complex of digoxin with hydroquinone [35] and benzocaine with caffeine were enhanced compared to the pure drugs [36].

SURFACE AREA OF THE DRUG PARTICLES

A decrease in particle size increases the specific surface area and, consequently, the dissolution rate and sometimes the solubility [37] of the drug. Traditional methods of decreasing particle size include fluid energy micronization, trituration and grinding, controlled precipitation by change of solvents or temperature, application of ultrasonic waves, ball milling, spray drying, administration of water-soluble salts of poorly soluble compounds from which the parent neutral forms precipitate in ultrafine form in the gastrointestinal fluids [38], freeze drying [39], solvation and desolvation in some cases [40], and by the method of solid dispersion discussed in detail in the following sections. The effect of particle-size reduction is a concomitant increase in the absorption of several drugs, e.g., proquazone [41], benoxaprofen [42], tolbutamide [43], griseofulvin [44], and tetracycline [45]. Kraml et al. [46] found appreciably higher blood levels from micronized griseofulvin compared to its regular crystalline form when administered orally to humans. However, in another study the administration of either 0.5 g of micronized or 1.0 g of regular crystalline griseofulvin to humans produced similar blood levels [47].

DEFINITION OF SOLID DISPERSIONS

It is logical here to define the term *solid dispersion* as applied in this book. The term refers to the dispersion of one or more active ingredients in an inert carrier or matrix at solid state prepared by the melting (fusion), solvent or the melting-solvent method." The dispersion of a drug or drugs in a solid diluent or diluents by traditional mechanical mixing is not included in this category. The term *coprecipitate* (more accurately *coevaporate*) has also been frequently used when a solid dispersion is prepared by the solvent method, such as coprecipitates of sulfathiazole-polyvinylpyrrolidone [48] and reserpine-polyvinylpyrrolidone [49]. However, the definition can now be broadened to include certain nanoparticles, microcapsules, microspheres, and other dispersions of drugs in polymers prepared by using any one of the above processes. Although most solid dispersion systems initially focused on producing increased dissolution rates, recent attempts have been aimed at achieving other goals such as sustained release of drugs, altered solid-state properties, enhanced release of drugs from ointment and suppository bases, and improved solubility and stability.

CLASSIFICATION OF SOLID DISPERSIONS

Solid dispersions have been classified mainly into six major categories as suggested by Chiou and Riegelmen [38], (1) simple eutectic mixtures, (2) solid solutions, (3) glass solutions of suspension, (4) compound or complex formations between the drug and the carrier, (5) amorphous precipitations of a drug in a crystalline carrier, and (6) any combinations of these groups.

SIMPLE EUTECTIC MIXTURES

An eutectic mixture of a sparingly water-soluble drug and a highly water-soluble carrier may be regarded thermodynamically as an intimately blended physical mixture of its two crystalline components. These components are assumed to crystallize simultaneously in very small particulate sizes. The increase in specific surface area, therefore, is mainly responsible for the increased rate of dissolution of a poorly water-soluble drug.

Differential thermal analysis (DTA) of binary mixtures normally exhibits two endotherms, but a binary mixture of eutectic composition usually exhibits a single major endotherm. In the case of a simple eutectic system, the thaw points of binary mixtures of varying compositions are equal to the eutectic temperature of the system.

Chiou and Niazi [50] examined griseofulvin-succinic acid solid dispersions prepared by the fusion method. They showed, with the aid of x-ray diffraction and DTA, that solid solubility was negligible, contrary to earlier suggestions, and classified this system as a simple eutectic mixture. Additional studies [51] of fused compositions of griseofulvin in succinic acid showed that the dissolution of griseofulvin was inversely proportional to the concentration of griseofulvin in the dispersion. This led to the conclusion that the increase in dissolution was mainly due to the decreased particle size obtained, although other factors such as increased wettability, reduction or absence of aggregation, and solubilization of the drug by the carrier at the site of the diffusion layer may have also contributed. On the other hand, Goldberg et al. [52] obtained similar dissolution profiles for both the fused solid dispersion of acetaminophen-urea at the eutectic composition and a physical mixture of the same composition.

SOLID SOLUTIONS

Solid solutions consist of a solid solute dissolved in a solid solvent. If the carrier is crystalline, a mixed crystal is formed because the two components crystallize together in a homogeneous one-phase system [38]. Perhaps, as suggested by Goldberg et al. [52], particle size is reduced in solid solution to molecular level, i.e., the dissolution of the drug occurs in the solid-state matrix. Hence, this system would be expected to yield much higher rates of dissolution than simple eutectic systems. In practice, the occurrence of solid solubility of less than 2% is considered insignificant. In binary systems, where solid solution formation is evident, the phase diagram is characterized by the disappearance of thaw points at a temperature higher than the eutectic temperature.

Chiou and Riegelman [53] reported a marked increase in dissolution rates of the sparingly water-soluble drugs digitioxin, 17-methyl testosterone, hydrocortisone acetate, and predinisolone acetate when dispersed in PEG 6000. This is believed to be due to formation of colloidal or molecular dispersion of the drug in the carrier. Similarly, Goldberg et al. [54] obtained a large increase in dissolution from the fused mixtures of chloramphenicol-urea, which formed a solid solution.

GLASS SOLUTION OF SUSPENSION

The principle of glass solution formation was first reported by Chiou and Riegelman [55] to enhance drug dissolution and absorption. A glass solution is a homogeneous system in which a glassy or a vitreous form of the carrier solubilizes drug molecules in its matrix. PVP dissolved in organic solvents undergoes a transition to a glassy state upon evaporation of the solvent. Glass solutions of digitoxin with PVP [56], methisazone with PVP [57], corticosteroids with dextrose or galactose [58], sulfamethoxazole with sugars [59], and primidone with citric acid [60] have been reported.

COMPOUND OR COMPLEX FORMATION

This system is characterized by complexation of two components in a binary system during solid dispersion preparation. The availability of a drug from the complex is dependent on the solubility, dissociation constant, and the intrinsic absorption rate of the complex. For example, PVP has been shown to retard the pharmacological actions of penicillin, novocaine, prostigmine, hexobarbital, quinine and hexylresorcinol [38]. Similarly, Geneidi et al. [61] reported a decrease in dissolution rate of nitrofurantoin from its coprecipitate or physical mixture with PVP 25,000 because of the formation of an insoluble complex.

AMORPHOUS PRECIPITATION

Amorphous precipitation occurs when the drug precipitates as an amorphous form in the inert carrier. The high energy state of the drug in this system generally produces much greater dissolution rates than the corresponding crystalline forms of the drug.

The conversion of a drug to an amorphous form of coprecipitation resulting in increased dissolution has been reported for sulfisoxazole-PVP [62] and chloramphenicol-PVP or -hydroxypropyl cellulose [63] systems. There may be other less clear examples of amorphous precipitation such as the phenytoin-PVP or sulfamethizole-PVP coprecipitate systems [64,65] where the incidence of coacervation of the carrier is believed to play a role in the particle size and forms of the released drug.

ULTRAMICROSIZE GRISEOFULVIN

Chiou and Riegelman [66] have shown that the oral absorption of griseofulvin was complete from its aqueous solution, only 88% from a 9:1 PEG 6000:griseofulvin solid dispersion, only 45% from a commercial capsule, and 33% from a tablet. It has been suggested that "ultramicrosize" griseofulvin is produced in the PEG 6000 matrix. It has been reported that the efficiency of oral absorption from ultramicrosize systems is approximately twice that of conventional micronized griseofulvin [67], but this is not always the case [68].

METHODS OF PREPARING SOLID DISPERSIONS

Generally, there are only two methods of preparing solid dispersions: fusion or solvent process. However, a third method of fusion-solvent may also be described.

FUSION PROCESS

In the fusion method of preparation, the carrier is heated to a temperature just above its melting point and the drug is incorporated into the matrix. The mixture is cooled with constant stirring to homogeneously disperse the drug throughout the matrix. Several mechanisms could operate during the process of dispersion. If the drug has a high degree of solubility in the carrier, the drug could remain "dissolved" in the solid state, yielding what is known as a solid solution. Particle-size reduction under these conditions proceeds to the ultimate level leading to molecular dispersion of the drug in the carrier matrix. These systems show very high drug dissolution rates compared to control samples. If, on the other hand, the solubility of the drug in solid state is not so high, crystallites of the drug become dispersed in the matrix. Such systems show only moderate increases in dissolution rates. A third mechanism is the conversion of a drug to an amorphous form in the presence of the matrix, again exhibiting different dissolution rates and solubility. Other factors that may play a role include solubilizing effect conferred by the carrier itself, improved wetting or decreased surface hydrophobicity, complexation, and crystallization of the drug in a metastable polymorphic form of altered thermodynamic properties. An important limitation of the fusion method of preparation is the exposure of drugs to elevated temperatures, particularly if the carrier is a high-melting solid and the drug is heat-sensitive.

Sekiguchi and Obi [69] used fusion to melt a sulphathiazole-urea mixture of eutectic composition above its eutectic temperature, solidified the dispersion on an ice bath, and pulverized it to a powder. Since a supersaturation of the drug can be obtained by quenching the melt rapidly (when the solute molecules are arrested in the solvent matrix by instantaneous solidification), rapid congealing is favored. Consequently, the solidification process is often effected on stainless steel plates [55,70] to favor rapid heat loss. A modification of the process involves spray congealing from a modified spray drier onto cold metal surfaces, which has been used for dispersions containing mannitol [71] or phenylbutazone-urea [72]. Spray-congealing processes are preferable since pellets of the dispersion can be produced without grinding and without altering the crystalline modification of the drug [73]. Walker et al. [74] demonstrated the feasibility of liquid-filling gelatin capsules with the liquid melt and avoiding grinding-induced changes in crystallinity.

The fusion process is technically the less difficult method of preparing dispersions, provided the drug and carrier are miscible in the molten state. Certain drug carriers, e.g., tolbutamide-mannitol [75], display a miscibility gap within their phase diagram, and consequent irregular crystallization may lead to only moderate increases in dissolution rate and difficulties in formulation. Kanig [71], as early as 1964, predicted that immiscibility and instability may occur during fusion, and Goldberg et al. [76] highlighted other potential problems such as thermal degradation, sublimation, and polymorphic transformation since metastable modifications of the drug may be formed, which convert to more stable forms during storage. Small crystallites may be obtained by quench cooling [77], but the solidification temperature will affect crystallization rates and may alter both the crystallite size and the hardness of the dispersion [78]. The solidified melt may be tacky and unhandable, and consequently novel formulation techniques are required to permit formulation into elegant dosage forms [74,79,80]. Some dispersions, e.g., griseofulvin-polyethylene glycol 6000, hardened on storage that favored pulverization, whereas others, e.g., griseofulvin-citric acid, require storage at elevated temperatures, e.g., 37°C, to facilitate hardening. However, this derivation of tacky or glassy dispersions or their comminution may induce crystallization and modify their dissolution characteristics [81].

Decomposition should be avoided during fusion, but is often composition-dependent [82,83] and affected by fusion time and the rate of cooling [84]. Therefore, to maintain decomposition at an acceptable level, fusion may be effected at a temperature just in excess of that which completely melts both drug and carrier, although it is feasible to prepare dispersions at just above the eutectic temperature when the carrier level is to the excess of the eutectic composition [80].

SOLVENT METHOD

In the solvent method of preparation, the carrier(s) and the active ingredient(s) are dissolved in a suitable organic solvent and the solvent is evaporated at an elevated temperature or under vacuum. As the solvent is being removed, supersaturation occurs followed by simultaneous precipitation of the constituents resulting in a solid residue. The coprecipitate is then dried under vacuum to drive out any solvent freely adhering to the particle surface. However, there is a possibility of the formation of a solvate with any of the constituents, which could hold some solvent within the crystal lattice. This presents a problem in terms of pharmaceutical acceptance since most of the solvents used are nonaqueous (organic) and toxic. Hence, removal of even trace amounts of the solvent is implied. Highly sensitive techniques such as differential scanning calorimetry (DSC), differential thermal analysis (DTA), thermogravimetric analysis (TGA), and less sensitive procedures like gravimetry and spectroscopy can be used to demonstrate complete solvent removal. The solvent process was initially used by Tachibana and Nakamura [85] when chloroform co-dissolved β-carotene and polyvinylpyrrolidone. Typically, large volumes of solvent were used since 500 mL chloroform was employed to dissolve 1 gram of solids. The dispersion was recovered by filcasting over mercury. More commonly, the solvent is removed by evaporation under reduced pressures at varying temperatures. For instance, Malone et al. [86] used a temperature of $-5°C$ and reduced pressure followed by drying for 12 hours in vacuum to remove ethanol, Kim and Jarowski [87] used 40°C to remove methanol/chloroform, and Chiou and Riegelman [66] either 115°C for 1 hour or 125°C for 25 minutes to remove ethanol from griseofulvin-PEG 6000 dispersions. Such variations in time and temperature merely indicate that complete removal of the solvent is often a long and difficult process. Stevens and Padfield [88] manipulated dispersions under petroleum ether at 60–80°C with a spatula to remove final traces of chloroform.

The choice of solvent and its removal rate are critical to the quality of dispersion. Since the chosen carriers are generally hydrophilic and the drugs are hydrophobic, the selection of a common solvent is difficult and its complete removal, necessitated by its toxic nature, is imperative. Certain solvents may plastisize polymeric carriers, e.g., polyvinylpyrrolidone (PVP), making their complete removal even more difficult [88]. Careful control of the temperature and rate of evaporation of solvents is essential in controlling the particle size of the drug, and although low temperature and vacuum evaporation may be used for solvent removal, instabilities cannot always be avoided. The tolbutamide-PVP dispersion showed an instability that varied to the evaporating temperature [89].

More rapid removal of the solvent may be accomplished by freeze drying. Polyvinylpyrrolidone dispersions of ketoprofen [90] or dicoumerol [91] were freeze dried from their ammoniacal solutions. Similarly, spray drying has been used to disperse acetohexamide in PVP [92] or chlorthalidone in pentaerythritol [93]. The difficulties in selecting a solvent common to both drug and carrier may be overcome by using an azeotropic mixture of the solvent in water [93].

The choice of the solvent may be further complicated by the required polymorphic form of the drug. Different polymorphs of spironolactone were obtained when its dispersions in PVP were co-evaporated from their solutions in ethanol, acetonitrile, or chloroform [94]. The ethanolic dispersion possessed the highest dissolution rates whereas the chloroformic evaporate gave the lowest.

Another disadvantage of the solvent method is that the volumes of solvents used may be excessive and the cost of their recovery prohibitive. For example, 500 mL of ethanol was used to prepare only 5 grams of 10% griseofulvin dispersion in polyethylene glycol 6000 [66].

Solid dispersion prepared by solvent removal processes were termed by Bates [49] as *coprecipitates* and by others [91] as *coevaporates*. The term *coprecipitate* will be used in this book to describe dispersions prepared by solvent removal. Since the larger increases in dissolution rates are generally obtained from dispersions containing the molecularly dispersed drug, the drug-to-carrier ratio is particularly important [95]. For instance, the acetohexamide-PVP 25,000 dispersions containing in excess of 70% polymer were amorphous and gave rapid drug release, but those containing less than 70% PVP were increasingly crystalline and gave only low dissolution rates [91].

Sekikawa et al. [95] quantified the mechanisms of dispersion formation by solvent removal using drug-PVP models. As evaporation proceeds, the drug concentrations reach and exceed solubility. The PVP inhibits drug crystallization, by maintaining the supersaturation, the degree of which continues to increase. If the PVP concentration is high enough to inhibit crystallization of the drug, the dispersion will appear from solution without crystallization, giving a dispersion containing amorphous drug. However, should the PVP concentration be too low, crystallization will not be inhibited and the dispersion would contain crystalline drug. Crystallization retardation is not only dependent on the drug-to-carrier ratio, but also on the physical properties of the drug molecule and the method of preparation. Sulfisoxazole:PVP dispersions of 1:3 were amorphous, but a 10:1 ratio contained crystalline drug. However, a 1:3 caffeine:PVP did not form a coprecipitate [95]. Amorphous hydroflumethazide-PVP systems were prepared by spray drying at lower PVP weight fractions than by other solvent removal processes and appeared to contain both amorphous hydroflumethazide and amorphous drug-PVP complexes whereas a coprecipitate did not [96].

FUSION-SOLVENT METHOD

In the fusion-solvent method, a carrier(s) is/are melted and the drug(s) is/are incorporated in the form of a solution. If the carrier is capable of holding a certain proportion of liquid yet maintaining its solid properties, and if the liquid is innocuous, the need for solvent removal is eliminated. Otherwise, this method faces the same criticism of solvent retention described before. This method is particularly useful for drugs that have high melting points or that are thermolabile. The feasibility of the method has been demonstrated for spironolactone and griseofulvin dispersions in polyethylene glycol 6000 (PEG 6000) [97]. Other researchers also adopted this method [98].

Although there are advantages and disadvantages associated with all these methods, the choice of a method of preparation could affect the intended purpose of solid dispersion formulations. Najib and Salem [99] have shown that ibuprofen dispersions prepared by the fusion method gave higher solubilities than those prepared by the solvent method. Similarly, higher dissolution rates were obtained for sulfamethoxazole-PEG [100] or griseofulvin-PEG [101] dispersions prepared by the fusion method as compared to coprecipitation. However, Jafari et al. [102] have reported comparable results from dispersions prepared by coprecipitation and fusion methods.

A reference guide for the most recent investigations of solid dispersion methods is listed in Table 2.1 and discussed elsewhere in the book.

TABLE 2.1. A Quick Reference Guide on the Investigation of Solid Dispersion Technology.

Drug	Conc. (%)	Carrier	Process	References
Griseofulvin	1–20	PEG; xylitol	Fusion	Sjokvist, 1988; 1989
	10	PEG/SAA	Solvent	Sjokvist, 1992; 1993
	10	PEG/SAA	Fusion	Alden, 1992
	10–50	PEG/SAA	Fusion	Veiga, 1993
	5–20	clay	Solvent	Takahashi, 1991
Nifedipine	3.8–50	PEG	Fusion	Save, 1992
	6.7–25	sucrose esters	Solvent	Ntawukulilyayo, 1993
	5	PEG/PC	Solvent	Law, 1992
	10	PEG/trilaurin	Fusion	Watanabe, 1993
1,4-Dihydro-pyridines phthalate		HPC/HPMC	Solvent	Watanabe, 1993
Oxodipine	4–80	PEG	Solvent/ fusion	Veiga, 1988
Hydrochlor-thiazide	50–67	PVP/HPMC	Solvent, fusion	Meshali, 1992
Tolbutamide	6.3	BCD	Solvent	Kedzierewicz, 1990
	33	PEG, BCD	Fusion	Kedzierewicz, 1993
Glibenclamide	—	PEG	Fusion	Kaali, 1992
Miconazole	—	PEG, PC	Fusion	Pedersen, 1990
Ketoconazole	—	PVP, MC	Solvent	Lee, 1991
Trimethoprim	5–60	PVP-XL	Phys. mixture	Giordano, 1988
Ciprofloxacin	10–50	PEG	Fusion	Francis, 1991
Naldixic acid	10,50	Modified starches	Solvent	Chowdary, 1992
Ibuprofen	40	Paraffin	Compressed discs	Healy, 1992
Ketoprofen	25–75	PC	Solvent	Fujii, 1988
	—	PEG	Solvent	Kim, 1991
Flurbiprofen	25–75	PC	Solvent	Fujii, 1988
	5	HPC	Solvent	Yuasa, 1993
Indomethacin	25–75	PC	Solvent	Fujii, 1988
	10,50	Eudragit E	Solvent	de Filippis, 1991
	10	PC	Solvent	Habib, 1993

TABLE 2.1. A Quick Reference Guide on the Investigation of Solid Dispersion Technology. (continued)

Drug	Conc. (%)	Carrier	Process	References
Piroxicam	10–80	PEG	Melting/ solvent	Fernandez, 1992; 1993
	—	PEG	Melting	Bhattacharyya, 1993
Mefenamic acid	—	Casein	Freeze drying	Suzuki, 1991
Phenyl- butazone	10,50	Modified starches	Solvent	Chowdary, 1992
Paracetamol	50	BCD; Ethocel Macrogol	Spray drying	Tasic, 1992
Benzo- diazepines	25–67	PC	Solvent	Fujii, 1991
Pheno- barbital	25–75	PC	Solvent	Fujii, 1991a
Nortriptyline HCl	10	PEG	Fusion	Craig, 1992
Oxazepam	9.1	PEG	Fusion	Jachowicz, 1993
	9.1	Collagel	Spray drying	Jachowicz, 1993
FK 506 (immuno- suppressant)	20	HPMC	Phys. mixt.	Kagayama, 1993
Methylparaben	40–60	CPG	Grinding	Yonemuchi, 1992
Terfenadine	5–40	PEG, PVP	Solvent	Badwan, 1991
α-Pentyl-3- (2-quino- linyl methoxy) benzene methanol	15–45	PEG, gelucire	Fusion	Serajuddin, 1988
	15–45	PEG/SAA	Fusion	Serajuddin, 1989
Albendazole	5–10	PVP	Solvent	Torrado, 1993

Note: The author gratefully acknowledges Dr. James A. Rogers, who prepared this table.

CHARACTERIZATION OF SOLID DISPERSIONS

A number of methods have been used to characterize solid dispersions including (1) thermal methods of analysis—differential thermal and thermomicroscopial; (2) powder x-ray diffraction; (3) microscopial studies, including the use of polarized light and the scanning electron microscope; (4) spectroscopic methods, especially I.R; (5) dissolution rate determination; (6) thermodynamic investigations involving determinations of the heats of dissolution, *H,* and the melting points in order to calculate the resulting changes in entropy; and (7) dynamic dialysis to characterize the formation of highly supersaturated solutions after dissolution of solid dispersions [38,103]. The most important and frequently used methods among these are thermoanalytical, powder x-ray diffraction and dissolution rate.

THERMOMICROSCOPICAL ANALYSIS

This is a visual method of analysis using a polarized microscope with a hot state to determine the thaw and melting points of solids. Its advantages are the small amount of sample required (virtually one crystal) and direct observation of the changes taking place in the sample through the thaw and melt stages. However, it does not provide the thermodynamics of the melting process and in some instances it is not as sensitive as DTA. The technique has been used by others often to support DTA or DSC measurements [104,105].

DIFFERENTIAL THERMAL ANALYSIS (DTA)

This is an effective thermal method for studying the phase equilibria of pure substances or solid mixtures. Differential heat changes that accompany physical and chemical changes are recorded as a function of temperature as the substance is heated at a uniform rate. In addition to thawing and melting, polymorphic transitions, evaporation, sublimation, desolvation, and other types of changes such as decomposition of the sample can be detected. DTA records energy changes occurring in the sample as it is being heated either exothermic or endothermic. However, for the interpretation of DTA thermograms, prior knowledge of the type of reactions that may be occurring is essential. For instance, it is necessary to know whether the sample is undergoing polymorphic change, decomposition, or desolvation. DTA has been used routinely to identify different types of solid dispersions [106,107].

It has been shown by Borchardt and Daniels [108] that the total heat of reaction, ΔH, is proportional to the area under the DTA peak as described by,

$$\Delta H = K \int_{t^1}^{t^2} \Delta T dT = KA \qquad (20)$$

where, K is a proportionality constant, ΔT is the temperature differential, A is the area under the DTA peak, and t is the time.

POWDER X-RAY DIFFRACTION

X-rays have been used in crystal structure studies in two different ways: (1) single crystal x-ray crystallography dealing with the determination of bond angles and interatomic distances, and (2) powder x-ray diffraction dealing with the study of crystal lattice parameters, where the x-ray diffraction intensity from a sample is measured as a function of the diffraction angles. Thus, changes in the diffraction pattern indicate changes in crystal structure. The relationship between the wavelength, λ, of the x-ray, the angle of diffraction, θ, and the distance between each set of atomic planes of crystal lattice, d, is given by Bragg's equation [109]:

$$M\lambda = 2d \sin \theta \qquad (21)$$

where m represents the order of diffraction.

X-ray diffraction spectra of simple eutectic systems show peaks of each crystalline component. Any change in the crystal lattice parameter will displace the diffraction peaks. Solids solutions exhibit a gradual shift in the positions of the diffraction lines with changes in composition. The lattice parameters of complexes are markedly different from those of pure components. Hence, the x-ray diffraction method can also be used in detecting complex formation. However, its major drawback has been the inability to differentiate between amorphous precipitation and molecular dispersion of the lattice parameter of the solvent component is unchanged. The techniques have been frequently used by researchers to characterize solid dispersions [81,110].

REFERENCES

1 Noyes, A. A. and Whitney, W. R. 1897. "The rate of solution of solid substances in their own solutions." *J. Am. Chem. Soc.*, 19:930–4.

2 Noyes, A. A. and Whitney, W. R. 1897. "Ueber die auflosungs geschwindigkeit von festen stoffen in ihren eigenen losungen." *Z. Physik. Chem.*, 23:684–92.

3 Brunner, L. and Tolloczko, S. 1900. "Uber die auflosungs geschwindigkeit fester korper." *Z. Physik. Chem.*, 35:283–290.

4 Nernst. W. 1904. "Theorie de reaktions geschwindigkeit in heterogenen systemen." *Z. Physik. Chem.*, 47:52–55.

5 Brunner, E. 1904. "Reaktions geschwindigkeit in heterogenen systemen." *Z. Physik. Chem.*, 47:56–102.

6 Wurster, D. E. and Taylor, P. W. 1965. "Dissolution rates." *J Pharm Sci.,* 54:169–75.

7 Hixon, A. W. and Crowell, J. H. 1931. "Dependence of reaction velocity upon surface and agitation. I. Theoretical consideration." *Ind. Eng. Chem.,* 23: 923–31.

8 Pederson, P. V. 1977. "New method for characterizing dissolution properties of dry powders." *J. Pharm. Sci.,* 66:761–6.

9 Nielsen, A. 1961. "Diffusion controlled growth of a moving sphere. The kinetics of crystal growth in a potassium perchlorate precipitation." *J. Phys. Chem.,* 65:46–9.

10 Higuchi, W. I. and Hiestand, E. N. 1963. "Dissolution rates of finely divided drug powders. I. Effect of a distribution of particle sizes in a diffusion-controlled process." *J. Pharm. Sci.,* 52:67–71.

11 Nelson, K. G. and Shah, A. C. 1975. "Convective diffusion model for a transport-controlled dissolution rate process." *J. Pharm. Sci.,* 64(4):610–4.

12 Dankwerts, P. V. 1951. "Significance of liquid-film coefficients in gas absorption." *Ind. Eng. Chem.,* 43:1460–7.

13 Goyan, J. E. 1965. "Dissolution rate studies III. Penetration model for describing dissolution of a multiparticulate system." *J. Pharm. Sci.,* 54:645–7.

14 Weibull, W. 1951. "A statistical distribution function of wide applicability." *J. Appl. Mech.,* 18:293–7.

15 Langenbucher, F. 1972. "Linearization of dissolution rate curves by the Weibull distribution." *J. Pharm. Pharmacol.,* 24(12):979–81.

16 Wagner, J. G. 1970. "Biopharmaceutics 20. Rate of dissolution in vitro and in vivo. Part V. Factors affecting rate of dissolution of drugs from tablets and capsules and interpretation of dissolution rate data from in vitro testing of tablets and capsules." *Drug Intell. Clin. Pharm.,* 4:132–4.

17 Raghunathan, Y. and Becker, C. H. 1968. "Spary-congealed formulations of sulfaethidole (SETD) and waxes for prolonged-release medication. Effect of modifiers." *J. Pharm. Sci.,* 57:1748–55.

18 Carstensen, J. T., Lai, T. Y. and Prasad, V. K. 1978. "USP dissolution IV: comparison of methods." *J. Pharm Sci.,* 67(9):1303–7.

19 Shah, A. C., Peot, C. B. and Ochs, J. F. 1973. "Design and evaluation of a rotating filter-stationary basket in vitro dissolution test apparatus. I. Fixed fluid volume system." *J. Pharm. Sci.,* 62(4):671–7.

20 Martin, A., Swarbrick, J. and Cammarata, A. Eds. 1983. "Physical pharmacy and drug product design." In *Physical Pharmacy,* A. Martin, J. Swarbrick, A. Cammarata, Eds. Philadelphia: Lea & Febiger, 3rd ed., 574–91.

21 Nelson, E., Knoechel, E. L., Hamlin, W. E. and Wagner, J. G. 1962. "Influence of the absorption rate of tolbutamide on the rate of decline of blood sugar levels in normal humans." *J. Pharm. Sci.,* 51:509-14.

22 Lin, S. L., Lachman, L., Swartz, C. J. and Heubner, C. F. 1972. "Preformulation investigation. I. Relation of salt forms and biological activity of an experimental antihypertensive." *J. Pharm. Sci.,* 61(9):1418–22.

23 Berge, S. M., Binghley, L. D. and Monkhouse, D. C. 1977. "Pharmaceutical salts." *J. Pharm. Sci.,* 66(1):1–19.

24 Levy, G. 1964. "Effect of dosage form on drug absorption. A frequent variable in clinical pharmacology." *Arch. Int. Pharmacodyn. Ther.,* 1(152):59–68.

25 Aguiar, A. J., Kro, Jr., J., Kinkel, A. W. and Samyn, J. C. 1967. "Effect of polymorphism on the absorption of chloramphenicol from chloramphenicol palmitate." *J. Pharm. Sci.,* 56(7):847–53.

26 Haleblian, J. and McCrone, W. 1969. "Pharmaceutical applications of polymorphism." *J. Pharm. Sci.,* 58(8):911–29.

27 Haleblien, J. 1975. "Characterization of habits and crystalline modification of solids and their pharmaceutical applications." *J. Pharm. Sci.,* 64(8):1269–88.

28 Shefter, E. and Higuchi, T. 1963. "Dissolution behavior of crystalline solvated and nonsolvated forms of some pharmaceuticals." *J. Pharm. Sci.,* 52:781–91.

29 Poole, J. W., Owen, G., Silverio, J., Freyhof, J. N. and Rosenman, S. B. 1968. "Physiochemical factors influencing the absorption of the anhydrous and trihydrate forms of ampicillin." *Curr Ther Res Clin Exp.,* 10(6):292–303.

30 Bates, T., Fung, H. L., Lee, H. and Tembo, A. V. 1975. "Comparative bioavailability of anhydrous griseofulvin and its chloroform solvate in man." *Res Commun Chem Pathol Pharmacol.,* 11(2):233–43.

31 Chiou, W. L. and Kyle, L. E. 1979. "Differential thermal, solubility, and aging studies on various sources of digoxin and digitoxin powder: biopharmaceutical implications." *J. Pharm. Sci.,* 68(10):1224–9.

32 Chow, K. Y., Go, J., Mehdizadeh, M. and Grant, D. J. W. 1984. "Modification of adipic acid crystals: influence of growth in the presence of fatty acid additives on crystal properties." *Int. J. Pharm.,* 20:3–24.

33 Burt, H. M. and Mitchell, A. G. 1981. "Crystal defects and dissolution." *Int. J. Pharm.,* 9:137–52.

34 York, P. and Grant, D. J. W. 1985. "A disruption index for qualifying the solid state disorder induced by additives or impurities. I. Definition and evaluation from heat of fusion." *Int. J. Pharm.,* 25:57–72.

35 Higuchi, T. and Ikeda, M. 1974. "Letter: Rapidly dissolving forms of digoxin: hydroquinone complex." *J. Pharm. Sci.,* 63(5):809–11.

36 Higuchi, W. I., Mir, N. A. and Desai, S. J. 1965. "Dissolution rates of polyphase mixtures." *J. Pharm. Sci.,* 54:1405–10.

37 Fincher, J. H., Adams, J. G. and Beal, H. M. 1965. "Effect of particle size on gastrointestinal absorption of sulfisoxazole in dogs." *J. Pharm. Sci.,* 54:704–8.

38 Chiou, W. L. and Riegelman, S. 1971. "Pharmaceutical applications of solid dispersion systems." *J. Pharm. Sci.,* 60(9):1281–302.

39 Suzuki, E., Shirotani, K., Tsuda, Y. and Sekiguchi, K. 1979. "Studies on methods of particle size reduction of medicinal compounds. VIII. Size reduction by freeze-drying and the influence of pharmaceutical adjuvants on the micromeritic properties of freeze-dried powders." *Chem. Pharm. Bull.* (Tokyo), 27(5):1214–22.

40 Sekiguchi, K., Ito, K., Owada, E. and Ueno, K. 1964. "Studies on the method of size reduction of medicinal compounds II. Size reduction of griseofulvin by solvation and desolvation method using chloroform (2)." *Chem. Pharm. Bull.* (Tokyo), 12:1192–7.

41 Nimmerfall, F. and Rosenthaler, J. 1980. "Dependence of area under the curve on proquazone particle size and in vitro dissolution rate." *J. Pharm. Sci.,* 69(5):605–7.

42 Ridolfo, A. S., Thompkins, L., Bechtol, L. T. and Carmichael, R. H. 1979. "Benoxaprofen, a new anti-inflammatory agent: particle-size effect on dissolution rate and oral absorption in humans." *J. Pharm. Sci.,* 68(7):850–2.

43 Nelson, E., Long, S. and Wagner, J. G. 1964. "Correlation of amount of metabolite excreted and its excretion rate with available surface area of Tolbutamide in dosage form." *J. Pharm. Sci.,* 53:1224–7.

44 Kraml, M., Dubuc, J. and Beall, D. 1962. "Gastrointestinal absorption of griseofulvin. I. Effect of particle size, addition of surfactants and corn oil on the level of griseofulvin in the serum of the rats." *Can. J. Biochem. Physiol.,* 40:1449–51.

45 Nelson, E. 1959. "Influence of dissolution rate and surface on tetracycline absorption." *J. Am. Pharm. Assoc.,* 48:96–103.

46 Kraml, M., Dubuc, J. and Dvornik, D. 1963. "Gastrointestinal absorption of griseofulvin. Levels in the serum of human patients on prolonged B.I.D. administration of micronized and common griseofulvin." *Arch. Derm.* (Chicago), 87:179–82.

47 Marvel, J. R., Schlichting, D. A., Denton, C., Levy, E. D. and Cahn, M. M. 1964. "The effect of a surfactant and of particle size on Griseofulvin plasma levels." *J. Invest. Dermatol.,* 42:197–203.

48 Simonelli, A. P., Mehta, S. C. and Higuchi, W. I. 1969. "Dissolution rates of high energy polyvinylpyrrolidone (PVP)-sulfathiazole coprecipitates." *J. Pharm. Sci.,* 58(5):538–49.

49 Bates, T. R. 1969. "Dissolution characteristics of reserpine-polyvinylpyrrolidone co-precipitates." *J. Pharm. Pharmacol.,* 21(10):710–2.

50 Chiou, W. L. and Niazi, S. 1973. "Differential thermal analysis and X-ray diffraction studies of griseofulvin-succinic acid solid dispersions." *J. Pharm. Sci.,* 62(3):498–501.

51 Chiou, W. L. and Niazi, S. 1976. "Pharmaceutical applications of solid dispersion systems: dissolution of griseofulvin-succinic acid eutectic mixture." *J. Pharm. Sci.,* 65(8):1212–4.

52 Goldberg, A. H., Gibaldi, M. and Kanig, J. L. 1966. "Increasing dissolution rates and gastrointestinal absorption of drugs via solid solutions and eutectic mixtures. II. Experimental evaluation of a eutectic mixture: urea-acetaminophen system." *J. Pharm. Sci.,* 55:482–7.

53 Chiou, W. L. and Riegelman, S. 1971. "Increased dissolution rates of water-insoluble cardiac glycosides and steroids via solid dispersions in polyethylene glycol 6000." *J. Pharm. Sci.,* 60(10):1569–71.

54 Goldberg, A. H. and Gibaldi, M. 1966. "Increasing dissolution rates and gastrointestinal absorption of drugs via solid solutions and eutectic mixtures. IV. Chloramphenicol—urea system." *J. Pharm. Sci.,* 55(6):581–3.

55 Chiou, W. L. and Riegelman, S. 1969. "Preparation and dissolution characteristics of several fast-release solid dispersions of griseofulvin." *J. Pharm. Sci.,* 58(12):1505–10.

56 Stupak, E. I. and Bates, T. R. 1973. "Enhanced absorption of digitoxin from orally administered digitoxin-polyvinylpyrrolidone coprecipitates." *J. Pharm. Sci.,* 62(11):1806–9.

57 Gidwani, R. N. and Anderson, A. J. 1976. "An in vitro evaluation of methisazone solid dispersions." *Can. J. Pharm. Sci.,* 11:117–20.

58 Allen, Jr., V. L., Yanchickn, V. A. and Maness, D. D. 1977. "Dissolution rates of corticosteroids utilizing sugar glass dispersions." *J. Pharm. Sci.,* 66(4):494–7.

59 Ghanem, A., Meshali, M. and Ibraheem, Y. 1980. "Dissolution rates of sulfamethoxazole utilizing sugar glass dispersions." *J. Pharm. Pharmacol.,* 32(10):675–7.

60 Summers, M. P. and Enever, R. P. 1976. "Preparation and properties of solid dispersion system containing citric acid and primidone." *J. Pharm. Sci.,* 65(11):1613–7.

61 Geneidi, A. S., Ali, A. A. and Salama, R. B. 1978. "Solid dispersions of nitrofurantoin, ethotoin, and coumarin with polyethylene glycol 6000 and their coprecipitates with providone 25,000." *J. Pharm. Sci.,* 67(1):114–6.

62 Sekikawa, H., Nakano, M. and Arita, T. 1978. "Dissolution behaviors and gastrointestinal absorption of sulfisoxazole in sulfisoxazole-polyvinylpyrrolidone coprecipitate." *Yakugaku Zasshi.,* 98(1):62–6.

63 Moriyama, M., Inoue, A., Isoya, M., Tanaka, M. and Hanano, M. 1978. "Dissolution properties and gastrointestinal absorption of chloramphenicol from hydrophilic high molecular compound coprecipitates." *Yakugaku Zasshi.,* 98(8):1012–8.

64 Sekikawa, H., Fujiwara, J., Naganuma, T., Nakano, M. and Arita, T. 1978. "Dissolution behaviors and gastrointestinal absorption of phenytoin in phenytoin-polyvinylpyrrolidone coprecipitate." *Chem. Pharm. Bull.* (Tokyo), 26(10): 3033–9.

65 Sekikawa, H., Nakano, M. and Arita, T. 1979. "Dissolution mechanisms of drug-polyvinylpyrrolidone coprecipitates in aqueous solution." *Chem. Pharm. Bull.* (Tokyo), 27(5):1223–30.

66 Chiou, W. I. and Riegelman, S. 1970. "Oral absorption of griseofulvin in dogs: increased absorption via solid dispersion in polyethylene glycol 6000." *J. Pharm. Sci.,* 59(7):937–42.

67 Barrett, W. E. and Hanigan, J. 1975. "The bioavailability of griseofulvin PEG ultramicrosize (Gris-PEG) tablets in man under steady-state conditions." *Curr. Ther. Res. Clin. Exp.,* 18(3):491–500.

68 Straughn, A. B., Meyer, M. C., Raghow, G. and Rotenberg, K. 1980. "Bioavailability of microsize and ultramicrosize griseofulvin products in man." *J. Pharmacokinet. Biopharm.,* 8(4):347–62.

69 Sekiguchi, K. and Obi, N. 1961. "Studies on absorption of eutectic mixtures. I. A comparison of the behavior of eutectic mixture of sulfathiazole and that of ordinary sulfathiazole in man." *Chem. Pharm. Bull.,* 9:866–72.

70 Sekiguchi, K., Obi, N. and Ueda, Y. 1964. "Studies on absorption of eutectic mixture. II. Absorption of fused conglomerates of chloramphenicol and urea in rabbits." *Chem. Pharm. Bull.* (Tokyo), 12:134–44.

71 Kanig, J. L. 1964. "Properties of fused mannitol in compressed tablets." *J. Pharm. Sci.,* 53:188–92.

72 Kreuschner, K., Fromming, K. H. and Hosemann, R. 1980. "Plug compounds—A new physical configuration of solid dispersions using the example of solidified phenylbutazone-urea melt embeddings." *Acta Pharm. Tech.,* 26:159–64.

73 Froemming, K. H., Simons, B., Haase, J. and Hosemann, R. 1978. "Problems of preparation, processing and storage of melt-embedded polyethylene glycol products." *Pharm. Ind.,* 40:967–70.

74 Walker, S. E., Ganley, J. A., Bedford, K. and Eaves, T. 1980. "The filling of molten and thixotropic formulations into hard gelatin capsules." *J. Pharm. Pharmacol.,* 32(6):389–93.

75 El-Banna, H. M., Daabis, N. A., Mortada, L. M. and Abd-Elfattah, S. 1975. "Physicochemical study of drug binary systems. Part 3: Tolbutamide-urea and tolbutamide-mannitol systems." *Pharmazie,* 30(12):788–92.

76 Goldberg, A. H., Gibaldi, M. and Kanig, J. L. 1965. "Increasing dissolution rates & gastrointestinal absorption of drugs in a solid solution and eutectic mixtures. I. Theoretical considerations and discussion of the literature." *J. Pharm. Sci.,* 54:1145–8.

77 Collett, J. H., Flood, B. L. and Sale, F. R. 1976. "Some factors influencing dissolution from salicylic acid-urea solid dispersions." *J. Pharm. Pharmacol.,* 28(4):305–8.

78 Daabis, N. A., Abd-Elfattah, S. and El-Banna, H. M. 1974. "Physicochemical study of drug binary systems. 2. Khellin-urea system." *Pharmazie,* 29(6): 400–4.

79 Ford, J. L. and Rubinstein, M. H. 1980. "Formulation and ageing of tablets prepared from indomethacin-polyethylene glycol 6000 solid dispersions." *Pharm. Acta Helv.,* 55(1):1–7.

80 Ford, J. L. and Rubinstein, M. H. 1981. "Preparation, properties and aging of tables prepared from the chlorpropamide-urea solid dispersion." *Int. J. Pharm.,* 8:311–22.

81 Chiou, W. L. 1977. "Pharmaceutical applications of solid dispersion systems: X-ray diffraction and aqueous solubility studies on griseofulvin-polyethylene glycol 6000 systems." *J. Pharm. Sci.,* 66(7):989–91.

82 Chiou, W. L. and Niazi, S. 1971. "Phase diagram and dissolution-rate studies on sulfathiazole-urea solid dispersions." *J. Pharm. Sci.,* 60(9):1333–8.

83 Ford, J. L., Stewart, A. F. and Rubinstein, M. H. 1979. "The assay and stability of chlorpropamide in solid dispersion with urea." *J. Pharm. Pharmacol.,* 31(11):726–9.

84 Carcamo, E. C. and Gana, I. M. 1974. "Eutectic mixtures and solid solutions of acetylsalicylic acid and urea. Stability of acetylsalicylic acid." *An. R. Acad. Farm.,* 40:487–93.

85 Tachibana, T. and Nakamura, N. 1965. "A method of preparing an aqueous colloidal dispersion of organic materials by using water-soluble polymers: dispersion of β-carotene by poly (vinyl pyrrolidone)." *Kolloid-Z.,* 203:130–3.

86 Malone, M. H., Hochman, H. I. and Nieforth, K. A. 1966. "Desoxycholic acid enhancement of orally administered reserpine." *J. Pharm. Sci.,* 55(9):972–4.

87 Kim, K. H. and Jarowski, C. I. 1977. "Surface tension lowering and dissolution rate of hydrocortisone from solid solutions of selected n-acyl esters of cholesterol." *J. Pharm. Sci.,* 66(11):1536–40.

88 Stevens, L. A. and Padfield, J. M. 1977. "Co-precipitated systems of salicylic acid (SSA) with polyvinyl pyrrolidone (PVP): physical and bioavailability studies." *Congr. Int. Technol. Pharm.,* 5:135–41.

89 Sekikawa, H., Naganuma, R., Fujiwara, J., Nakano, N. and Arita, T. 1971. "Dissolution behaviors and gastrointestinal absorption of tolbutamide in tolbutamide-polyvinyl pyrrolidone coprecipitate." *Chem. Pharm. Bull.* (Tokyo), 27(1):31–7.

90 Takayama, K., Nambu, N. and Nakai, T. 1982. "Factors affecting the dissolution of ketoprofen from solid dispersions in various water-soluble polymers." *Chem. Pharm. Bull.,* 30:3013–6.

91 Sekikawa, H., Fukuda, W., Takada, M., Ohtani, K., Arita, T. and Nakano, N. 1983. "Dissolution behavior and gastrointestinal absorption of dicumarol from solid dispersion systems of dicumarol-polyvinylpyrrolidone and dicumarol-beta-cyclodextrin." *Chem. Pharm. Bull.* (Tokyo), 31(4):1350–6.

92 Graf, E., Beyer, C. and Abdallah, O. 1982. "Solid dispersions of acetohex-amide. 3. Acetohexamide-poly (vinyl pyrrolidinone). Methods of preparation and solvents used." *Acta Pharm. Tech.,* 28:225–30.

93 Bloch, D. W., Elegakey, M. A. and Speiser, P. 1983. "Spray-dried solid dis-persions of hydrochlorothiazide and chlorthalidone in pentaerythritol." *Pharm. Acta Helv.,* 58:14–22.

94 El-Dalsh, S. S., El-Sayed, A. A., Badawi, A. A. and Fouli, A. 1982. "Effect of the solvent used in coprecipitation on the dissolution rate of spironolactone." *Pharmazie,* 37:606–7.

95 Sekikawa, H., Nakanoa, M. and Arita, T. 1978. "Inhibitory effect of polyvinylpyrrolidone on the crystallization of drugs." *Chem. Pharm. Bull.,* 26:118–26.

96 Corrigan, O. I., Sabra, K. and Holohan, E. M. 1983. "Physicochemical prop-erties of spray-dried drugs: Phenobarbitone and hydroflumethiazide." *Drug Dev. Ind. Pharm.,* 9:1–20.

97 Chiou, W. I. and Riegelman, S. 1971. "Pharmaceutical applications of solid dispersion systems." *J. Pharm. Sci.,* 60(9):1281–302.

98 Fernandez, M., Rodriguez, I. C., Margarit, M. V. and Cerezo, A. 1992. "Char-acterization of solid dispersions of piroxicam/polyethylene glycol 4000." *Int. J. Pharm.,* 84:197–202.

99 Najib, N. M. and Salem, M. A. S. 1987. "Release of ibuprofen from polyeth-ylene glycol solid dispersions: Equilibrium solubility approach." *Drug Del. Ind. Pharm.,* 13:2263–75.

100 Singla, A. K. and Vijan, T. 1990. "Dissolution of sulfamethoxazole from poly-ethylene glycols and polyvinyl pyrrolidone solid dispersions." *Drug Dev. Ind. Pharm.,* 16:875–82.

101 Sjokvist, E. and Nystrom, C. 1988. "Physicochemical aspects of drug release. VI. Drug dissolution rate from solid particulate dispersions and the importance of carrier and dry particle properties." *Int. J. Pharm.,* 47:51–66.

102 Jafari, M. R., Danti, A. G. and Ahmed, I. 1988. "Comparison of polyethylene glycol, polyvinylpyrrolidone and urea as excipients for solid dispersion sys-tems of miconazole nitrate." *Int. J. Pharm.,* 48:207–15.

103 Kreuter, J. 1983. "Solid dispersion and solid solution." In *Topics in Pharma-ceutical Sciences,* D.D. Breimer, P. Speiser, Eds. Amsterdam, Netherlands: Elsevier Science Publishers, 359–69.

104 Ford, J. and Rubinstein, M. H. 1978. "Phase equilibria and dissolution rates of indomethacin-polyethylene glycol 6000 solid dispersions." *Pharm Acta Helv.,* 53(11):327–32.

105 Daabis, N. A., Abd-Elfattah, S. and El-Banna, H. M. 1974. "Physicochemical study of drug binary systems. 2. Khellin-urea system." *Pharmazie,* 29(6):400–4.

106 El-Banna, H. M., Abd-Elfattah, S. and Daabis, N. A. 1974. "Physicochemical study of drug binary systems. 1. Phenobarbital-urea system." *Pharmazie,* 29(6):396–400.

107 Rogers, J. A. and Anderson, A. J. 1982. "Physical characteristics and dissolu-tion profile of ketoprofen-urea solid dispersions." *Acta Pharm. Helv.,* 57:276–81.

108 Borchart, H. J. and Daniels, F. 1957. "The application of differential thermal analysis to the study of reaction kinetics." *J. Am. Chem. Soc.,* 79:41–6.

109 Willard, H. H., Merritt, Jr., L. L. and Dean, J. A. 1965. "X-Ray methods." In *Instrumental Methods of Analysis.* New York: Van Nostrand Reinhold Company, 198–238.

110 McGinity, J. W., Maincent, P. and Steinfink, H. 1984. "Crystallinity and dissolution rate of tolbutamide solid dispersions prepared by the melt method." *J. Pharm. Sci.,* 73(10):1441–4.

Please return to:

**PHARMA
R & D LIBRARY**

CHAPTER 3

Phospholipids as Tools to Improve Drug Dissolution

GOPI K. VUDATHALA[1]

INTRODUCTION

SYSTEMIC absorption of most drug products consists of a succession of rate processes, which include (1) disintegration of the drug product and subsequent release of the drug; (2) dissolution of the drug in an aqueous environment; and (3) absorption across cell membranes into the systemic circulation. The rate at which the drug reaches the blood circulation is determined by the slowest step in the sequence, which can also be termed the *rate-limiting* step. Except for sustained-release or prolonged-action products, disintegration of a solid dosage form occurs prior to drug dissolution and drug absorption. The rate at which poorly water-soluble drugs dissolve is often the slowest step and therefore exerts a rate-limiting effect on drug bioavailability [1].

In the case of drugs with dissolution rate-limited absorption, reduction of particle size often increases the rate of dissolution and amount of drug absorbed. For example, the therapeutic dose of griseofulvin was reduced to 50% by micronization, and a more constant and reliable blood level was produced [2]. More sophisticated means of physically reducing particle size has been achieved by solid dispersion formulation [3]. Solvent deposition onto inert solid particles has also increased the surface area available for dissolution [4]. Solid dispersions have been investigated for many years. The main cate-

[1]Procter & Gamble Pharmaceuticals, Cincinnati, OH, U.S.A.

gories include: (1) eutectics, (2) amorphous precipitates, (3) solid solutions and (4) glass solutions.

Dramatic increases in dissolution of drugs have been shown from these types of formulations. For example, a 900-fold increase in the dissolution rate of chlorpropamide from dispersions of 30% w/w in urea has been reported [5]. There have also been several reports of improved dissolution of griseofulvin (Gris) from solid dispersions with inert adjuvants, such as polyvinylpyrrolidone (PVP) [6], polyethylene glycol (PEG) [7)], polyoxyethylene 40 monostearate (POS) [8], pentaerythritol [9], and succinic acid [10]. The mechanisms of increased dissolution rates of solid dispersions of drugs in PVP carriers have been reviewed by Simonelli [11]. The main advantages of using water-soluble polymers as carriers are their nontoxicity and general applicability to most drugs [3]. In spite of the many apparent advantages of solid dispersions, only two products are commercially available, namely, Nabilone-PVP (Cesamet) and Griseofulvin-PEG (Gris-PEG). The main problems in commercialization of these formulations are their instabilities during processing and in long-term storage. Some studies have identified and quantified these problems but attempts to reduce or eliminate them have not been addressed.

DISSOLUTION THEORY

Frequently, dissolution is the rate-limiting or rate-controlling step in the bioabsorption of drugs of low solubility. The rate at which a solid dissolves in a solvent was proposed quantitatively by Noyes and Whitney in 1897 [12]. They stated that the rate at which a solid substance dissolves in its own solution is proportional to the difference between the concentration of the saturated solution and the concentration of that solution, and is expressed mathematically as,

$$\frac{dC}{dt} = k\,(C_s - C) \tag{1}$$

where C is the concentration at time t, and C_s is the equilibrium solubility of the solute at the experimental temperature, dC/dt is the dissolution rate, and k is a proportionality constant. The model used assumed that a thin layer of saturated solution is formed initially at the crystal-liquid interface, that the rate of dissolution is governed by the rate of diffusion from this layer to the bulk of the solution, and that there is negligible change in the surface area with time during dissolution. The surface area, S was then incorporated into Equation (1) to give

$$\frac{dC}{dt} = k_1 S \left(C_s - C \right) \tag{2}$$

where k_1 is the intrinsic dissolution rate constant. Brunner and Nernst [13] applied Fick's law of diffusion to establish a relationship between the constant (k_1) in Equation (2) and the diffusion coefficient of the solute, thus transforming Equation (1) to:

$$\frac{dC}{dt} = \frac{DS}{Vh} \left(C_s - C \right) \tag{3}$$

where D is the diffusion coefficient of the solute, V is the volume of the dissolution medium, and h is the thickness of the diffusion layer. This has been referred to as the Nernst-Brunner film theory of dissolution. Other dissolution theories have been reviewed extensively by Higuchi [14], Wagner [15] and Leeson and Carstensen [16].

Hixson and Crowell in 1931 derived an equation for the dissolution of drug powders consisting of uniformly sized particles that expresses the rate of dissolution based on the cube root weight of particles (Hixson-Crowell cube root law) [17]. Other dissolution models of interest that have been developed include Danckwert's model, the convective-diffusion model [18], the Higuchi-Heistand model [19], and the multiparticulate dissolution model [20].

PHOSPHOLIPIDS AS EXCIPIENTS IN SOLID DISPERSIONS

The effect of lipids on drug dissolution in the GI tract has been referred to some extent previously [21]. However, there are reports on the specific application of phospholipid (lecithin) to alter the dissolution behavior of poorly water-soluble drugs. Specifically, improved bioavailability of griseofulvin in rats was observed after oral administration of a griseofulvin suspension containing 0.5% lecithin

[22]. The improved dissolution [23], release characteristics [24], and bioavailability of griseofulvin from griseofulvin-phospholipid co-precipitate systems have been demonstrated [25]. Recently, the improved solubilities of indomethacin, ketoprofen, and flurbiprofen have been reported from (1:2 mole ratio) coprecipitates with phosphatidylcholine [26]. Fujii et al. also reported significant increases in the dissolution behavior and rat plasma concentrations of (1:3 mole ratio) phenytoin-phosphatidylcholine solid dispersions [27]. These effects were considered to have occurred as a result of the existence of amorphous drug in phosphatidylcholine. The dissolution of griseofulvin from griseofulvin-hydrogenated soya phospholipid coprecipitates was reported to be accelerated as a result of a decrease in crystallinity of griseofulvin and possible aggregation of phospholipid with griseofulvin [28].

Phospholipids are one of the major structural components of cell membranes. A comprehensive list of the distribution of phospholipids in cell membranes is available [29]. Phosphatidylcholine (PC) was first isolated from egg yolk and brain by Gobley in 1850.

The chemical name of phosphatidylcholine is: 1, 2-diacyl-sn-glycero-3-phosphocholine. Its molecular structure is represented as shown with fatty acyl substituents in Figure 3.1. In phosphatidylethanolamine and phosphatidylserine, the choline moiety is replaced by ethanolamine and serine, respectively. Other phospholipids that occur in tissues include phosphatidylethanolamine (PE), phosphatidylserine (PS), and phosphatidylglycerol (PG). Lysolecithin is a significant component of blood plasma but occurs in tissues at much lower levels than lecithin. Most naturally occurring lecithins contain both a saturated fatty acid and an unsaturated fatty acid, but there are notable exceptions [29]. Bilayers composed of a single phospholipid species undergo a well-defined thermotropic phase transition (in the fully hydrated state) in which the lipid chains change from an ordered gel state to a fluid or liquid crystalline state. The temperature at which this occurs is called the *phase transition temperature, T_c*. Determination of the T_c is composition but each depends strongly on the water content, with increasing water possible using DTA or DSC. The transition temperatures and phase behavior differ with content, the transition temperature decreases progressively, reaching a limiting value at 25–30 wt% water [30].

Also, the phase transition entropy and enthalpy are found to depend approximately linearly on the fatty acid chain length. Below

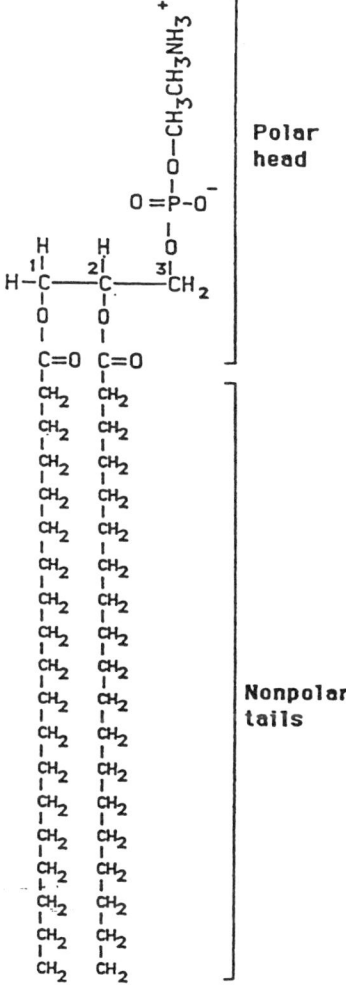

Phosphatidyl choline

Figure 3.1 Structure of phosphatidylcholine showing the polar and nonpolar regions. Reprinted with permission from *J. Pharm. Sci.*, 79(5):464.

the phase transition temperature, the fatty acyl side chains are in a closely packed, relatively ordered, extended all-trans conformation giving rise to the "gel state." As the temperature rises, the packing of the fatty acyl chains decreases and over a short temperature range (1–2°), the molecules become highly disordered (T_c). As the temperature is increased above the T_c, the phospholipid molecules again assume a regularly ordered arrangement and have the properties of a liquid-crystalline state. The extent of the molecular motions can be detected by electron spin resonance (ESR) or nuclear magnetic resonance (NMR) probes and the fluidity can be monitored by fluorescence depolarization analysis using a suitable fluorescent probe. The T_c is a function of the nature of the fatty acyl chains and the polar head group and can vary between -20 and $+90°C$.

Phospholipids have seldom been used in solid dosage forms, although it has been known for some time that under appropriate conditions, they are able to spontaneously form liposomes from a solid film and entrap polar or nonpolar molecules. Phospholipids and liposomes have potential use in drug delivery systems to provide sustained release of drugs, localized drug delivery, and enhanced uptake of drugs by target cells [25].

PREPARATION OF DRUG: PHOSPHOLIPID SOLID DISPERSIONS

Broadly, solid dispersions are prepared by either fusion or solvent methods. Phospholipids, due to their nature, cannot withstand high temperatures, and thus the fusion method will not yield the desired results unless the drug component has a low melting point. Most commonly, solvent methods are used to prepare drug:phospholipid solid dispersions. This involves dissolving the phospholipid and the drug component in an organic solvent in a jacketed beaker and evaporating the solvent at relatively low temperatures with constant stirring and/or a gentle stream of nitrogen. The coprecipitate formed is vacuum dried overnight and then scraped off the beaker. The resulting powder is passed through a desired mesh-size sieve to obtain a uniform powder.

DISSOLUTION OF DRUG: PHOSPHOLIPID SOLID DISPERSIONS

Recent studies have demonstrated the potential of improving the solubility, dissolution, and bioavailability of griseofulvin from griseofulvin-phospholipid coprecipitate systems [25]. The dissolution rate of griseofulvin from coprecipitates containing as little as 5% DMPC was 3-fold greater than that of micronized or solvated griseofulvin or its physical mixture in pH 2.0 HCl-KCl buffer at 37°C. The dissolution profiles shown in Figure 3.2(a) describe an initial rapid dissolution rate followed by a gradual leveling off after about 20 to 40 minutes. It is apparent from these data that Gris underwent increased dissolution from coprecipitates that contained DMPC compared to the pure drug, but compositions that included CHOL in the lipid component at a 1:0.33 mole ratio caused the dissolution of Gris to approximately double. However, as the CHOL content increased, the fraction of Gris dissolved after 120 minutes decreased, and Gris:CHOL coprecipitates at a 4:1 weight ratio exhibited dissolution behavior similar to that of pure or solvated Gris, although the initial dissolution rate (5 min) was slightly greater.

No eutectic or solid solution formation in these coprecipitates was indicated from the phase diagram. X-ray diffraction spectra indicated the presence of griseofulvin crystallites in the coprecipitates. The existence of griseofulvin in the chloroform-solvated state was deemed essential for enhancement of dissolution. The improved dissolution rate from coprecipitates has been thought to occur from metastable drug-phospholipid-chloroform crystals as a result of the rapid hydration of lecithin at the crystal surfaces resulting in rapid, extensive disintegration into extremely fine particles of drug and formation of myelinic structures at the crystal surfaces, which are thought to contain entrapped griseofulvin (Figure 3.3). The result is an effective increase in the saturation concentration of drug in the theoretical diffusion layer during the dissolution process [Equation (3)] [23]. However, the performance of the coprecipitates was found to be affected by processing and aging of the coprecipitates. Thus, aged coprecipitates were found to undergo a slower rate of dissolution than fresh samples. The oral administration of freshly prepared coprecipitates to rats yielded a significant increase in the relative bioavailability of griseofulvin, which correlated with the dissolution behavior [31,32]. When part of the phospholipid (in the coprecipitates) was replaced with cholesterol, the initial dissolution rates decreased whereas the fraction dissolved after 60 minutes increased. Consequently, it was suggested that adjustment of the cholesterol content could be a means of controlling the rate of release of griseofulvin from these systems [24].

The dissolution of other drug solvates, including the steroids, may also benefit by formulation with phospholipids. Preliminary tests

44

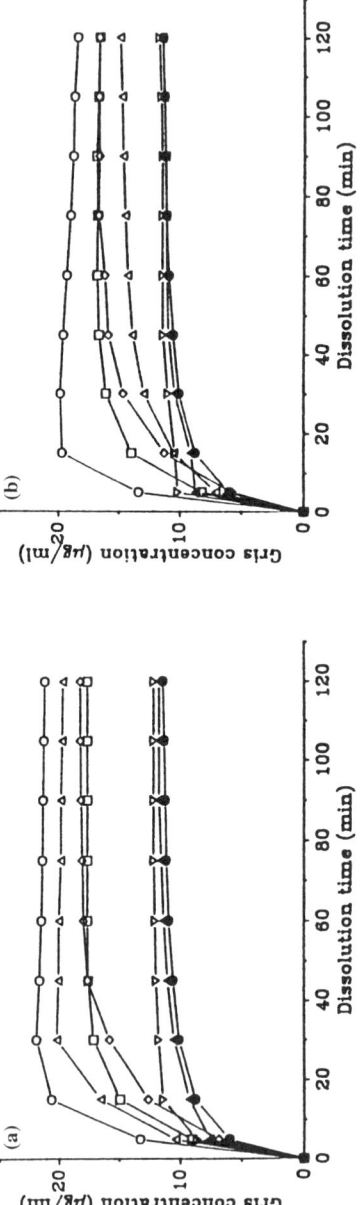

Figure 3.2 (a) Dissolution of Gris:DMPC:CHOL coprecipitates after aging for 1 day. (△) 4:1 (1:0); (○) 4:1 (1:0.33); (□) 4:1 (1:1); (◇) 4:1 91:3); (▽) 4:1 (0:1); (▲) Chloroform-solvated Gris; (●) micronized Gris. The weight ratio Gris:lipid was constant but the mole ratio DMPC:CHOL (in parentheses) was varied in the coprecipitates. (b) Dissolution of Gris:DMPC:CHOL coprecipitates after aging for 90 days. (△) 4:1 (1:0); (○) 4:1 (1:0.33); (□) 4:1 (1:1); (◇) 4:1 91:3); (▽) 4:1 (0:1); (▲) Chloroform-solvated Gris; (●) micronized Gris. The weight ratio Gris:lipid was constant but the mole ratio DMPC:CHOL (in parentheses) was varied in the coprecipitates. Reprinted with permission from *Int. J. Pharmaceutics*, 69:13–19, 1991.

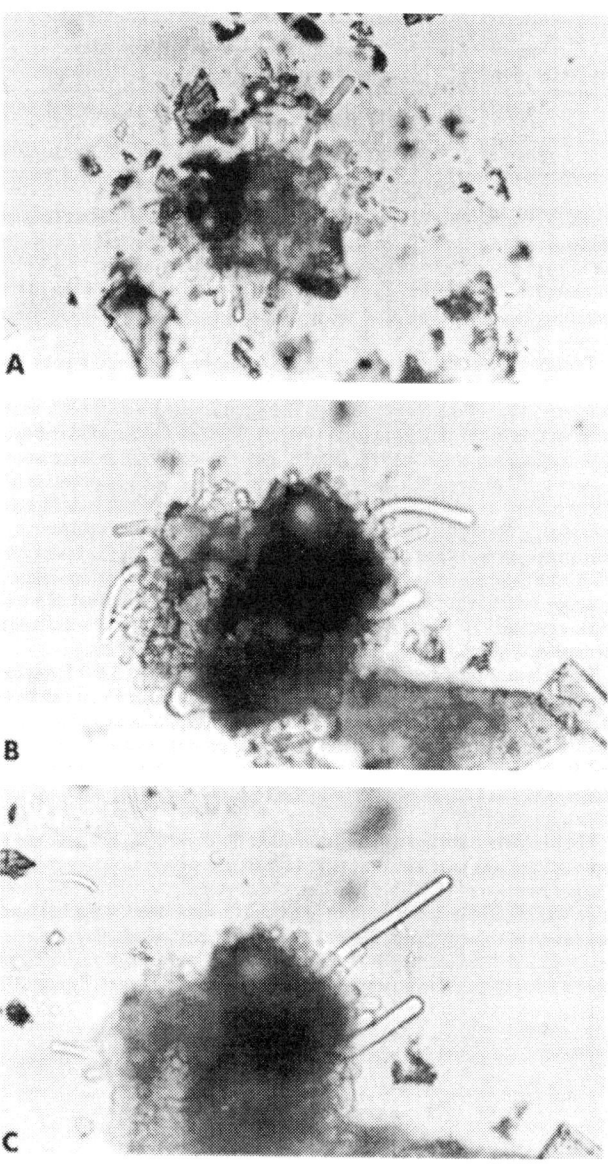

Figure 3.3 Photomicrographs of micronized griseofulvin:DMPC (5:1) coprecipitate crystals dispersed in water at various time periods after preparation (320 × magnification). Key: (a) 2 minutes; (b) 3 minutes; (c) 5 minutes. Reprinted with permission from *J. Pharm. Sci.* 73(6):757, 1984.

demonstrated that the dissolution of hydrocortisone from its di-
methylformamide solvate increased significantly after incorporation
of only 5% dimyris-toylphosphatidylcholine [25]. Fludrocortisone
acetate forms a pentanol or an ethyl acetate solvate [33] and its dissolution from these solvates has been reported [34]. Although fludrocortisone acetate exhibits improved dissolution from its solvate,
it is poorly soluble in water (40 mg/L) [35]. Furthermore, since it
would be beneficial not only to have increased dissolution properties but also controlled release behavior, the effect of incorporation
of selected polymers in the coprecipitates on the release of the drug
was determined [32]. A comparison of the relative dissolution behavior of FA alone, as a solvate, or as a coprecipitate with DMPC
in a dissolution medium of pH 2.0 at 37°C is shown in Figure 3.4.
The fludrocortisone acetate (FA) or FA-solvate dissolved rapidly
then leveled off at about 28 µg/mL although the FA-solvate had
considerably greater dissolution properties over the first 10–15 minutes. The maximum concentration of FA achieved in solution is lower
than the reported solubility of FA (40 mg/L), which is probably due
to the low pH of the medium. In contrast, a marked improvement in
dissolution of FA from DMPC coprecipitates was observed and is
consistent with previous observations involving griseofulvin [23].

Recent studies have demonstrated significant improvement in dissolution of other poorly water-soluble drugs, such as indomethacin

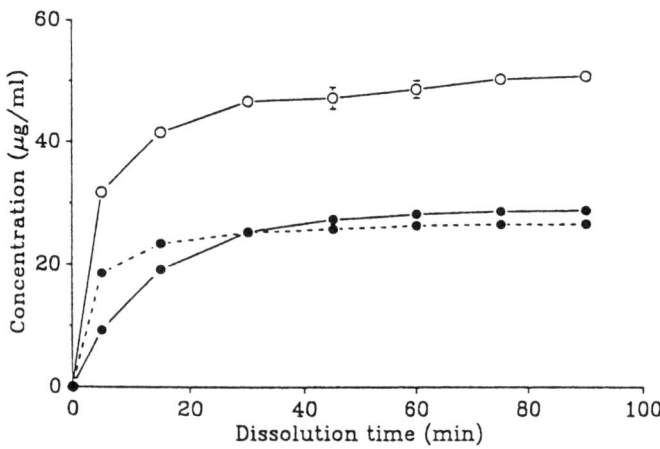

Figure 3.4 Dissolution profiles of (-●-) FA, (- - ● - -) FA-solvate, and (○) FA:DMPC (4:1,
w/w) coprecipitate at pH 2.0 and 37°C. Reprinted with permission from *J. Pharm. Sci.,* 81:
282–286, 1992.

(IND) [36] and carbamazepine (CBZ) [37] from their phospholipid solid dispersions. It is apparent from Table 3.1 that the initial dissolution rate (IDR, computed over the first five minutes of dissolution) and the total amount dissolved after 60 minutes from the coprecipitates exceeded those of pure IND or the corresponding physical mixture. The table shows a quantitative measure of the increased dissolution with the IND-DMPG solid dispersion. The IDR of the coprecipitate exhibited a 6-fold increase with a small (6%) amount of DMPG in the total sample (16:1 weight ratio) when examined as a fresh sample and a greater than 5-fold increase when examined after one year of preparation [36]. Different phospholipids show various degrees of improvement in the dissolution profiles (Figure 3.5). In this case, DSPC and DMPG:CHOL (1:1) show less improvement in the release of the drug than DMPG alone. Similarly, a 10:1 ratio of CBZ to phospholipid yielded a greater than 3-fold increase in IDR than the pure CBZ (Figure 3.6) [37]. Also, the total amount dissolved after 60 minutes was more than twice greater as seen in the drug:phospholipid solid dispersion powders and also in finished dosage forms prepared from the blends of the solid dispersion with other excipients. For example, capsule formulations of IND coprecipitates yielded a 2-fold greater IDR as well as total amount dissolved in 60 minutes, compared to marketed capsules (Indocin®, Merck Sharp & Dohme) (Figure 3.7). Further, CBZ:DMPG (10:1) coprecipitates when formulated into tablets showed a 2-fold increase in dissolution (IDR and amount dissolved in 60 minutes) compared to a commercially available tablet of CBZ (Tegretol®) (Figure 3.8). Powder x-ray diffraction spectra showed no changes in the diffraction patterns of CBZ in the coprecipitate. CBZ is sparingly soluble in water and has shown erratic bioavailability charac-

TABLE 3.1. Dissolution of IND-DMPG Composition at 37°C and in Distilled Water

Composition (weight ratio)	Initial Dissolution Rate (mg/100 mL/min)	Concentration After 60 min (mg/100 mL)
1:0 (pure IND)	0.036	1.8
16:1 (physical mixt)	0.040	2.0
16:1 (fresh coppt)	0.217	4.3
16:1 (1-year-old coppt)	0.197	4.0

Reprinted with permission from *Drug Dev. Ind. Pharm.*, 19:1225, 1993.

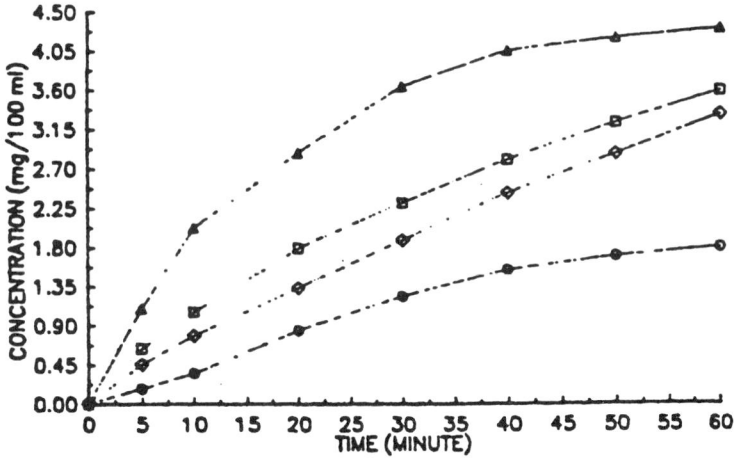

Figure 3.5 Dissolution behavior of IND:PL solid dispersion system at 37°C and in distilled water. Key: (O) pure IND; coprecipitates at a 16:1 ratio of IND:PL − ( ) DSPC; (~) DMPG:CHOL (1:1); (ε) DMPG. Reprinted with permission from *Drug Dev. Ind. Pharm.*, 19:1224, 1993.

teristics after oral dosing. It also has a narrow therapeutic index even after controlled administration. The rate of CBZ release absorption is controlled by the release of drug from its dosage form in the gastrointestinal tract. Therefore, modifications in the dissolution profile of CBZ have obvious clinical significance, and thus CBZ:PL solid dispersions may have clinical advantages of quick *in vivo* release to

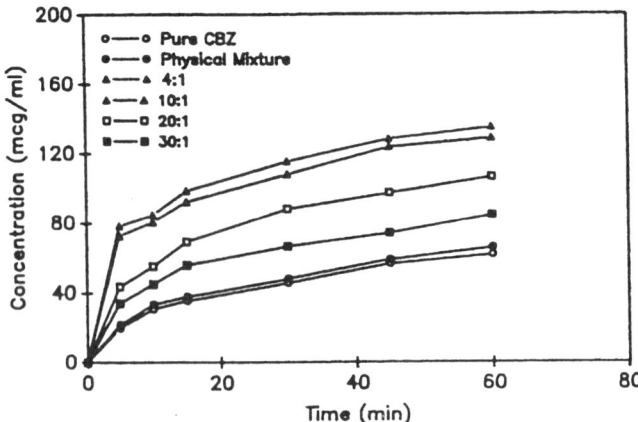

Figure 3.6 Dissolution profile of carbamazepine from phospholipid solid dispersions prepared at various ratios of carbamazepine:DMPG. Reproduced with permission from *J. Contr. Rel.*, 23:242, 1993.

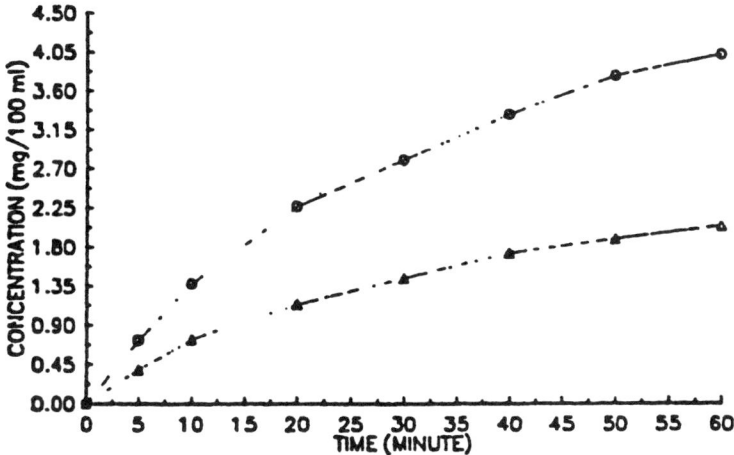

Figure 3.7 Dissolution of IND solid dispersion capsules at 37°C and in distilled water. Key: (○) IND:DMPG (16:1) capsule; (△) Indocin® capsule. Reprinted with permission from *Drug Dev. Ind. Pharm.*, 19:1227, 1993.

yield better bioavailability than existing commercial formulations. This concept may apply to several poorly water-soluble drugs.

Data from some of the above studies are summarized in Table 3.2. It is clear that solvates of drugs generally exhibit slower rates of dissolution than the pure drug forms. Drug:phospholipid (PL) solid dispersions with as little as 6% of phospholipid content exhibit increased dissolution by more than 2-fold compared to that of the pure

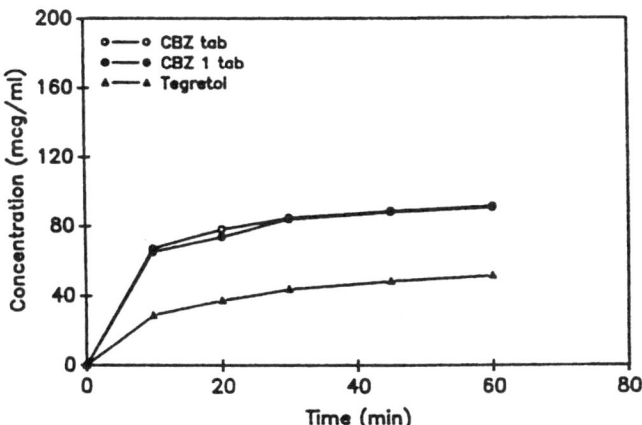

Figure 3.8 Comparison of tablet dissolution rate. Reprinted with permission from *J. Contr. Rel.*, 23:244, 1993.

TABLE 3.2. Dissolution Comparisons of Various Drug Forms (% Dissolved in Respective Media after 60 Minutes).

Drug	Pure Form	Solvate	Drug: PL Solid Dispersion	Medium	Dosage Form of Pure Drug	Dosage Form of Drug: PL Solid Dispersion
Griseofulvin	29	25	56 (4:1)[a]	pH 2.0 HCl-KCl buffer	23.5	45
Indomethacin	27	—	64 (16:1)	Distilled water	30	63
Carbamazepine	28	—	61 (4:1)	Distilled water	20	40
Fludrocortisone acetate	50	48	92 (4:1)	pH 2.0 HCl-KCl buffer	—	—

[a]Drug: PL ratio.

drug. This behavior is also seen with the dosage forms (tablets or capsules) (Figure 3.5) containing drug:phospholipid solid dispersions where a greater than 2-fold increase in dissolution is observed in comparison with the dosage form manufactured using the pure drug form.

MECHANISM OF DISSOLUTION

The dissolution of a drug from a solid dispersion system consists of three processes: (1) the coprecipitate interacts with water in its vicinity, (2) finely dispersed drug in the matrix is released, and (3) solubilized drug is supersaturated in the diffusion layer. Some of the most common reasons for increased dissolution rates from solid dispersions have been the formation of solid solution, eutectics, conversion to an amorphous form, solubilization effect by the carrier, or increased hydrophilicity of the drug due to coating by polymers. According to the film theory of dissolution, factors affecting the rate of dissolution include the diffusion coefficient, the thickness of diffusion layer, the surface area, and the difference between the saturation concentration in the diffusion layer and the concentration of bulk solution. Under identical conditions of dissolution testing for the same compositions of drug:phospholipid, differences in dissolution behavior between coprecipitates and physical mixtures or solvated drug are probably closely related to inherent differences in particle size.

Heating solvated griseofulvin can lead to desolvation and conversion to very small particles of anhydrous griseofulvin and in the process a 100% increase in surface area is realized. Cheng et al. [38], in a detailed study, reported that the desolvation process requires substantial reorganization of the crystals. This results in many regions, which nucleate separately and form small particles. It appears that in coprecipitate crystals a phospholipid alters the crystal structure in such a manner as to cause fracture and disintegration upon its spontaneous hydration after making contact with water. The evidence of this comes from microscopical observations of the coprecipitates in which very fine crystallites of griseofulvin were found to be released as the original crystals fractured in the dissolution medium. When physical mixtures are used or when coprecipitates are prepared from other solvents, the phospholipid undergoes dispersion in a similar manner but it does not bring about crystal disruption because it is not intimately involved in the crystal lattice. Hence, no increase in dissolution rate or extent is realized from these systems. The solubilization effect in the diffusion layer surrounding the particles may be exerting a substantial effect in the dissolution process. More specifically, it is the rapid, spontaneous dispersion of DMPC into colloidal aggregates of phospholipid bilayers or liposomes and the simultaneous incorporation of griseofulvin molecules into this partitioning system that is analogous to an increase in the saturation concentration in the diffusion layer. In the case of physical mixtures, DMPC is not in a suitable physical state for spontaneous dispersion and, in addition, griseofulvin molecules must be dissolved from the solid state before becoming sequestered by the preformed liposomes, a process that in itself is relatively slow [31].

Microscopic observation of coprecipitate crystals undergoing dissolution in water revealed multitudes of long finger-like structures forming at the crystal surfaces, growing and extending outward into the dissolution medium as part of a rapid dynamic process of dispersion and what may be considered as part of the dissolution process (Figure 3.3). A similar observation was made upon examination of the physical mixture although in this case dissolution was not improved. Although these microscopic observations were made in isolation in what may be considered a static system, they demonstrate an important phenomenon of phospholipid solid dispersion formulations, which under appropriate conditions can lead to improved dissolution [31].

Thus, phospholipids improve the dissolution of a drug by their ability to form liposomes, which increases the intrinsic solubility of the drug in the medium due to a partitioning process.

EFFECT OF PHOSPHOLIPID COMPOSITION ON DISSOLUTION

The release of drug from a drug:phospholipid solid dispersion varies depending on the type of phospholipid used. Figures 3.2 shows the effects of various phospholipids incorporated into griseofulvin and CBZ coprecipitates, respectively. It is apparent that the initial dissolution rate and the limiting concentration after 60 minutes are greater for all coprecipitates than those obtained from the pure or solvated drug. Furthermore, the release of drug decreases with an increase in the chain lengths of fatty ester of the phospholipid. For example, DSPC has a higher chain length than DMPC or EPC. Thus, DSPC shows less improvement in the release of the drug than DMPC or EPC. The increase in the release of drug may be due to an intrinsic solubility of the drug in the medium of the stationary layer or due to a partitioning process. If it is assumed that EPC, DPPC and DSPC behave like DMPC in the formation of the crystal lattice of the coprecipitate, the decreased release of griseofulvin, when combined with a phospholipid having a longer chain, is likely determined by the relative ease of dispersion of the phospholipid upon contact with the aqueous phase, which in turn is related to its T_c and the corresponding physical state at 37°C. The T_c of EPC, DMPC, DPPC and DSPC are 1°, 23°, 41°, and 58°C, respectively. Since the T_c of DPPC and DSPC are too far above the temperature of the dissolution medium (37°C), they would exist in a gel state at the experimental temperature and thus show only a minimal influence on the dissolution of the drug. Therefore, increasing the amount of DSPC in the coprecipitate from 19:1 to 4:1 makes little difference in the release of griseofulvin because it does not readily disperse below its T_c. On the other hand, EPC and DMPC show vastly improved dissolution rate and extent. DMPG demonstrates the most improvement in dissolution of CBZ. This may be due to the net negative charge on its polar head group, which could form an electrostatic complex with the amino group of CBZ. This favorable orientation could have increased CBZ concentration in the vicinity of DMPG liposomes and hence increased the initial rate of dissolution as observed.

Hence, the effect of phospholipids on dissolution of drugs appears to be related to their individual abilities to rapidly and spontaneously undergo dispersion in aqueous media. In this regard, it is known that DPPE alone does not readily disperse in water and, consequently, it does not influence the dissolution of griseofulvin from coprecipitates.

EFFECT OF COPRECIPITATING SOLVENT

The solvent used to prepare the coprecipitate sometimes binds to the drug crystals forming solvates. Solvate formation has in some cases contributed to increased dissolution, while in other cases it has led to decreased dissolution. Griseofulvin has been known to form solvates with chloroform, but not with ethanol or methylene chloride. Methylene chloride-treated griseofulvin has a slightly increased dissolution, while ethanol-treated griseofulvin has slightly decreased dissolution when compared to untreated or chloroform-treated griseofulvin. However, the various solvent-treated coprecipitate systems did not significantly alter the dissolution of griseofulvin except when chloroform was employed as the coprecipitating solvent (Figure 3.9). This phenomenon was not observed with CBZ precipitates. Chloroform, methanol, and ethanol yielded almost identical rates and extents of CBZ dissolution from CBZ:DMPG (10:1) coprecipitates (Table 3.3). Thus it can be concluded that the coprecipitating sol-

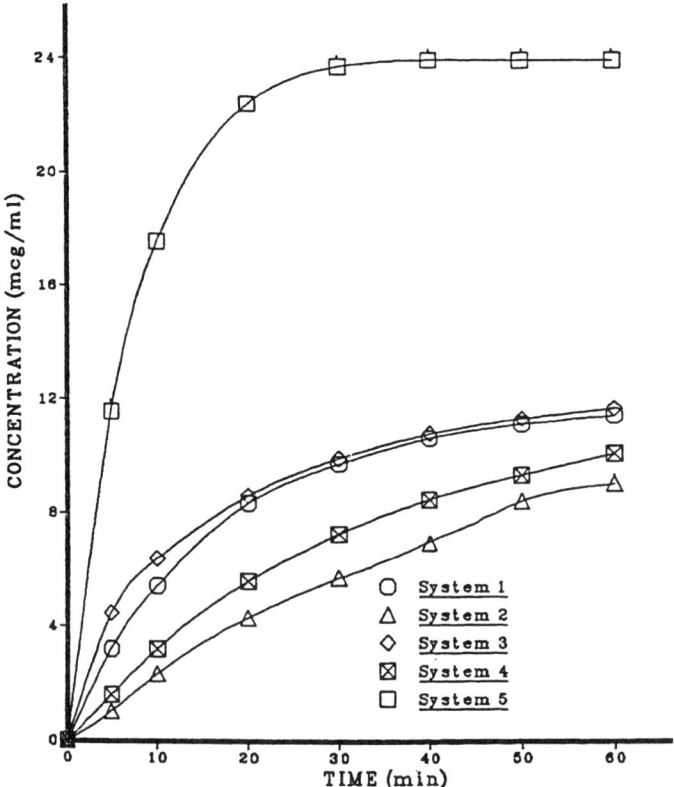

Figure 3.9 Effect of solvent selection on the dissolution of solvent-treated griseofulvin and griseofulvin:DMPC coprecipitates (4:1 weight ratio) in pH 2.0 HCL-KCl buffer at 37°C. System 1 and 2—griseofulvin treated with methylene chloride and ethanol respectively; systems 3, 4 and 5—griseofulvin:DMPC (4:1) coprecipitated with methylene chloride, ethanol, and chloroform, respectively. Reprinted with permission from Ph.D. thesis of Suresh Venkataram, University of Alberta, Canada, 1986.

TABLE 3.3. Comparison of Various Coprecipitating Solvents on the Dissolution of CBZ:DMPG (10:1) in Distilled Water and at 37°C.

Solvent	Amount Dissolved (:g/mL) in			
	5 min	15 min	30 min	60 min
Chloroform	72.8	92.3	108.0	128.8
Methanol	71.2	91.9	107.2	125.2
Ethanol	70.1	90.3	105.5	122.7

Reprinted with permission from *J. Contr. Rel.*, 23:243, 1993.

vent shows a pronounced effect on drug dissolution from certain drug:phospholipid coprecitates, especially when they involve drugs that form solvates, while it has no effects on drugs that do not form solvates. It should be noted that upon aging, the loss of the bound solvent to the atmosphere can result in decreased dissolution.

APPLICATION OF DRUG: PHOSPHOLIPID COPRECIPITATES IN CONTROLLED RELEASE

The application of microcapsule and microsphere technology has produced better physical properties of drugs for manufacturing tablets or capsules because of superior flow characteristics and compressibilities compared to the standard granulation. In addition, the microparticles processed in this manner can be formulated to yield drug-release kinetics that are more uniform and prolonged. Another potential use of microencapsulation in the formation of small particles is believed to be the stabilization of solid dispersions, thereby increasing the utility of these dispersions both after processing and during storage. Microencapsulated coprecipitates of the griseofulvin:phospholipid system, which were shown to have superior dissolution behavior compared to micronized griseofulvin, was attempted with moderate success in obtaining controlled release [39]. The release of Gris from Gris:phospholipid coprecipitates exhibited similar behavior in microencapsulated and nonmicroencapsulated states. In comparison, the release profile of Gris from microencapsulated Gris was 20% less than that obtained from its dissolution. Furthermore, the integrity of coprecipitates in microspheres suspended in PEG 600 for one week appeared to be maintained. The release of

Gris from coprecipitates incorporated in PLA microspheres was found to be similar to the dissolution of the coprecipitate, whereas the release of Gris was slowed down when incorporated in microspheres. Decreasing the molecular weight of PLA significantly increased the release of Gris from microencapsulated coprecipitates but not pure Gris. The release of Gris was only slightly reduced by its incorporation in 146,000 MW compared to 50,000 MW PLA microspheres, but the coprecipitate microspheres yielded considerably different release kinetics after the initial rapid-release phase, indicating the possibility of preparing controlled drug delivery systems of coprecipitates by varying the molecular weight of the polymer. Thus, microencapsulation of solid dispersion systems offers new opportunities for their application in improving the bioavailability of poorly water-soluble drugs and for the development of this type of formulation as pharmaceutical products.

STABILITY

Although the solid dispersion appears to be a potentially useful dosage form to increase dissolution and absorption rates of poorly water-soluble drugs, unfortunately, it is often physically unstable during storage under various conditions. The aging effects of eutectic mixtures, solid solutions, glass solutions, and metastable polymorphic forms have been presented somewhat in detail by Chiou and Riegelman [40]. More recently, Ford [41] discussed the disadvantages of solid dispersion systems. The main reasons for changes in the characteristics of the solid dispersions in storage were crystallization, hardening, self-association of the drug, brittleness, aggregation, and hygroscopicity. Aging had not been considered a problem in studies on drug-polyethylene glycol solid dispersions and recent studies on freshly prepared and aged tolbutamide-polyethylene glycol solid dispersions appear to support this [42]. However, aging may be important for some drug-polyethylene glycol systems but not others, and further studies are required to establish the same.

Pure urea inclusion complexes are relatively stable in the solid state. They can be stored at a relative humidity of 70–75% without noticeable deterioration. Tablets of urea inclusion compounds prepared with water-free additives have been reported to be stable for some years [43]. But in mixtures with water-containing additives, the urea complexes have a reduced stability due to autoxidation. Thus, manufacturing of the solid complex is only possible with water-free excipients. The stability behavior of solid inclusion compounds is different from that in solution. The cyclodextrin complexes can exist in the dissolved form, in which case an inhibitory or accelerating effect can be exerted on chemical reactions of the included guest compound. An inhibitory effect on the hydrolysis of benzocaine [44] and hydrolysis of digoxin [45] was demonstrated. On the other hand, higher humidity and higher water contents of additives in solid formulations have had no negative influences on cyclodextrin inclusion compounds compared to urea adducts. Kedzierewicz et al. [46] found that the x-ray diffraction patterns did not change after storage of the tolbutamide-β cyclodextrin complexes at room temperature for four months, with respect to the fresh complex. This observation indicates a good stability of the complex for a duration of at least four months. Further, it was also determined that the photoxidation of clofibrate can be inhibited if the drug is included as guest molecule in an inclusion compound with β- and Γ-cyclodextrins [45]. Possible stabilization of vitamin D_3 in the solid complex against heat, light, and oxygen has been reported [47]. The formulation was found to retain 94 percent of its original activity after storage in an uncovered glass container at 60°C for up to seven days.

The drug:phospholipid coprecipitates have also been shown to age as observed from the gradual reduction in the rate and extent of dissolution with time [25]. Thus, the dissolution profile of griseofulvin from griseofulvin:dimyristoylphosphatidyl-choline (DMPC) coprecipitates was similar to that of griseofulvin alone after aging for four months at room temperature. X-ray diffraction evidence of the aged sample demonstrated conversion to a crystalline structure not unlike that of pure Gris. Desolvation of the chloroform-solvated Gris in the coprecipitate was established as being the cause of the decreased dissolution in the aged sample [23]. Recently, it was discovered that the addition of cholesterol (CHOL) to certain compositions of Griseofulvin:DMPC coprecipitates increased the dissolution of griseofulvin, reduced the residual chloroform content, and decreased the aging towards lower dissolution behavior [48]. Coprecipitates of griseofulvin:lipid, 4:1 weight ratio (1:0:33 DMPC:CHOL mole ratio), exhibited maximum dissolution and the lowest degree of aging [Figure 3.1(b)]. Since the coprecipitates contain chloroform bound within the crystals, which participates in the crystalline structure together with phospholipid, the loss of chloroform with time is expected to be a factor in the aging of the coprecipitation. A comparison of the fusion temperatures, heats of fusion, and chloroform contents revealed that the 4:1 (1:0:33) composition underwent the least change of these parameters with age. From a plot of the aging coefficient, it was concluded that the introduction of small amounts of cholesterol to the lipid component beyond a threshold level, but over a narrow range, increased the storage stability of the precipitates. The replacement of DMPC with egg phosphatidylcholine (EPC), at the optimum mole ratio of 1:0.33 EPC:CHOL, resulted in increased dissolution of griseofulvin upon aging up to 90 days, the level of which remained constant for samples aged up to 135 days. This unexpected behavior was probably a result of the degradation of EPC. Another study determined that fludrocortisone acetate: DMPC coprecipitates did not age over a period of four months at room temperature [39], as a result of negligible losses of ethylacetate from the coprecipitates.

Therefore, finding different methods for preventing aging in solid dispersions can alleviate some of the stability problems associated with these systems and would contribute to increased commercialization of solid dispersions.

Poorly water-soluble drugs often have poor bioavailabilities due to low and erratic absorption. Those drugs that undergo dissolution rate-limited gastrointestinal absorption generally obtain improved dissolution and bioavailability by reduction of particle size. For example, the therapeutic dose of micronized griseofulvin was able to be reduced to 50%; additional benefits included steady and predictable blood levels. However, micronization of drugs often leads to aggregation and agglomeration of particles resulting in poor wettabilities. Solid dispersions of poorly water-soluble drugs with water-soluble carriers have reduced the incidence of these problems and improved dissolution [40].

One of the earliest solid dispersion systems prepared to enhance dissolution rates and absorption behavior of a drug was the solidified urea melt of griseofulvin. Subsequently, it has been established that variations in dissolution of drugs from solid dispersions are not only dependent on the water-soluble carrier but also on the preparation technique. Unfortunately, many solid dispersions are physically unstable due to the high surface energy of the particles produced and, consequently, have a short shelf-life. It is mainly for this reason that few commercial solid dispersion products are available. Solid dispersion systems that improved the dissolution of solvated griseofulvin, but also were subject to aging, were those formulated containing phospholipids [31]. Recently, it has been reported that the aging of these coprecipitates could be drastically reduced or even halted by adding a critical amount of cholesterol [48]. The bioavailabilities of griseofulvin:dimyristoylphosphatidylcholine coprecipitates in rats and *in vitro-in vivo* correlations have been reported by Venkataram and Rogers [31]. Extension of these studies to griseofulvin coprecipitate formulations that contain other phospholipids, cholesterol, and that compares the bioavailabilities of 1-day versus 90-day aged coprecipitates concluded that the advantageous effects of cholesterol addition on the dissolution behavior of Gris in Gris:PL coprecipitates are equally applicable to the *in vivo* situation. In a comparison of 1-day- and 90-day-aged Gris:DMPC (4:1w/w) coprecipitates (formulations B vs. G, Table 4), there were significant differences in C_{max}, AUC and t_{max} as expected from the *in vitro* studies. In contrast, formulations C and H (Gris:DMPC:CHOL [4:1(1:0.33)] coprecipitates) correlated with the 1-day-aged *in vitro* data (Table 3.4) but no differences were found statistically when formulation C was compared with formulation H using the *in vivo* parameters in Table 3.5. This is interpreted to mean that the reduction

TABLE 3.4. Pharmacokinetic Parameters after Oral Administration of Gris Formulations.[a]

Formulation[b]	C_{1h}, :g/mL	C_{max}, :g/mL	t_{max}, h	AUC, :g/mL.h[c]
1-Day-Aged				
A	0.39±0.13	1.32±0.19	5.40±0.80	12.11±0.91
B	1.13±0.12	1.83±0.10	2.20±0.40	11.16±0.94
C	1.62±0.60	1.97±0.76	1.80±0.75	13.55±4.84
D	0.98±0.19	1.23±0.16	1.60±0.49	7.04±1.45
90-Day-Aged				
G	0.99±0.49	1.19±0.64	1.60±0.49	7.37±3.54
H	1.49±0.29	1.60±0.39	1.40±0.49	9.90±2.56
I	1.31±0.52	1.72±0.10	1.60±0.80	9.17±2.75

[a]Data are reported as mean ±SD (n = 5).
[b]A, micronized Gris; B and G, Gris:DMPC (4:1 w/w) coprecipitate; C and H, Gris:DMPC:CHOL [4:1(1:0.33)] coprecipitate; D and I, Gris:EPC:CHOL [4:1(1:0.33)] coprecipitate.
[c]For 24 h after oral administration.
Reproduced with permission from *J. Pharm. Sci.*, 81:1169, 1992.

TABLE 3.5. Statistical Analysis of *in vivo* Parameters Obtained for Various Gris Formulations.[a]

Formulation Comparison[b]	C_{1h}	C_{max}	t_{max}	AUC
A vs. B	*	*	*	—
A vs. C	*	—	*	—
A vs. D	*	—	*	*
B vs. G	—	*c	*c	*c
C vs. H	—	—	—	—
D vs. I	—	*	—	—

[a]Asterisk (*) indicates significantly different at $p < 0.05$ unless otherwise indicated; —, not significantly different. Data were determined by the paired t test.
[b]See Table 3.4, footnote[a].
[c]Significantly different at $p < 0.1$.

in solvent loss from the coprecipitates containing 33% CHOL in the lipid phase minimized the aging process and, consequently, contributed to maintaining the improved dissolution behavior and absorption efficiency of these formulations. Comparison of formulations D and I revealed a significant increase in C_{max} with concomitant increases in C_{1h} and AUC. This apparently unusual increase in bioavailability with aging of the Gris:EPC:CHOL coprecipitates is attributed to degradation of EPC [48]. However, the exact mechanisms responsible for the increased dissolution and bioavailability of Formulation I are not completely understood.

CONCLUSIONS

Solid dispersions of poorly water-soluble drugs, such as griseofulvin, indomethacin, carbamazepine, fludrocortisone acetate, etc., prepared with a phospholipid improves the dissolution of these drugs, and the aging characteristics of the solid dispersions can be considerably slowed down by the incorporation of CHOL. The quantities of phospholipid used in drug:phospholipid solid dispersions are very small (5–20%) in comparison to other solid dispersion systems that have been developed with water-soluble components, e.g., polyethylene glycol, polyvinyl pyrrolidone (5–10% drug content). This provides for smaller dosage form unit weight, particularly for high-dose drugs, thus providing for patient convenience/ease of swallowing. Further, the dissolution behavior of dosage forms formulated with drug:phospholipid solid dispersions has shown promise in comparison to that with pure drugs, offering the possibilities of improved bioavailability, better stability, and reduced recall situations attributed to dissolution failures. Furthermore, the choice of phospholipid can make a remarkable difference in the dissolution and bioavailability outcome. This has further implications for other drugs that are poorly water-soluble and exhibit low dissolution and bioavailability. The aging process of the drug:phospholipid solid dispersions has proved not to be detrimental to achieving improved bioavailability in comparison with other solid dispersions of drugs with water-soluble carriers.

Upon contact with water, the crystals of the drug spontaneously disperse into extremely fine particles, followed by hydration of the phospholipid, which rapidly forms myelinic structures at the crystal surfaces. This process, along with partitioning of drug in the myelinic structures, accounts for the rapid initial dissolution rate of the drug and for the severalfold increase in the amount of drug dissolved over a period of 1–2 hours.

Microencapsulation of intact drug:phospholipid solid dispersion systems in microspheres has been shown to be possible. The release of Gris from Gris:phospholipid solid dispersions exhibited similar behavior in microencapsulated and nonmicroencapsulated states. Furthermore, the integrity of drug:phospholipid solid dispersions in microspheres suspended in PEG 600 for one week appeared to be maintained. It appears that microencapsulation may offer new opportunities to exploit the solid dispersion formulation advantages of poorly water-soluble drugs. This could entail minimizing the difficulties encountered with these systems during processing as well as developing desirable controlled-release patterns.

Finally, drug:phospholipid solid dispersions offer the potential to provide improved formulations of poorly water-soluble drugs with improved dissolution, bioavailability, and stability.

REFERENCES

1 Shargel L. Yu ABC. 1985. "Biopharmaceutic aspects of drug products," In *Applied Biopharmaceutics and Pharmacokinetics,* 2nd ed. East Norwalk, CT: Appleton Century-Crofts, p. 67.

2 Atkinson, R. M., Bedform, C., Child, K. J. and Tomich, E. G. 1962. "Human blood griseofulvin levels from different dosage schedules." *Antibiot. Chemother.* (NY), 12:225–31.

3 Chiou, W. L. and Riegelman, S. 1971. "Absorption characteristics of solid dispersed and micronized griseofulvin in man." *J. Pharm. Sci.,* 60(9):1376–80.

4 Yang, K. Y., Glemza, R., Jarowski, C. I. 1979. "Effects of amorphous silicon dioxides on drug dissolution." *J. Pharm. Sci.,* 68:560–5.

5 Ford, J. L. and Rubinstein, M. H. 1977. "The effect of composition and ageing on the dissolution rates of chlorpropamide-urea solid dispersions." *J. Pharm. Pharmacol.,* 29(11):688–94.

6 Mayersohn, M. and Gibaldi, M. 1966. "New method of solid-state dispersion for increasing dissolution rates." *J. Pharm. Sci.,* 55(11):1323–4.

7 Chiou, W. L. and Riegelman, S. 1970. "Oral absorption of griseofulvin in dogs: increased absorption via solid dispersion in polyethylene glycol 6000." *J. Pharm. Sci.,* 59(7):937–42.

8 Kaur, R., Grant, D. J. W. and Eaves, T. 1980. "Comparison of polyethylene glycol and polyoxyethylene stearate as excipients for solid dispersion systems of griseofulvin and tolbutamide II: dissolution and solubility studies." *J. Pharm. Sci.,* 69(11):1321–6.

9 Chiou, W. L. and Riegelman, S. 1969. "Preparation and dissolution characteristics of several fast-release solid dispersions of griseofulvin." *J. Pharm. Sci.,* 58(12):1505–10.

10 Goldberg, A. H., Gibaldi, M. and Kanig, J. L. 1966. "Increasing dissolution rates and gastrointestinal absorption of drugs via solid solutions and eutectic mixtures. IV. Chloramphenicol—urea system." *J. Pharm. Sci.,* 55(6):581–3.

11 Simonelli, A. P. 1969. "Dissolution rates of high energy polyvinylpyrrolidone (PVP)-sulfathiazole coprecipitates." *J. Pharm. Sci.,* 58(5):538–49.

12 Noyes, A. A. and Whitney, W. R. 1897. "The rate of solution of solid substances in their own solutions." *J. Am. Chem. Soc.,* 19:930–4.

13 Nernst, W. 1904. "Theorie der reaktions geschwindigkeit in heterogenen systemen." *Z. Physik. Chem.,* 47:52–5.

14 Higuchi, W. I. 1967. "Diffusional models useful in biopharmaceutics. Drug release rate process." *J. Pharm. Sci.,* 56:315–24.

15 Wagner, J. G. 1971. "Introduction to rate of dissolution *in vitro* and *in vivo.*" In *Biopharmaceutics and Relevant Pharmacokinetics.* Hamilton, IL: Drug Intelligence, pp. 98–103.

16 Leeson, L. J. and Carstensen, J. T. 1974. *Dissolution Technology.* Washington, D.C.: Academy of Pharmaceutical Sciences.

17 Martin, A. N., Swarbrick, J. and Cammarata, A. 1983. "Diffusion and dissolution." In *Physical Pharmacy: Physical Chemical Principles in the Pharmaceutical Sciences,* 3rd ed. Philadelphia: Lea & Febiger, pp. 399–444.

18 Nielsen, A. 1961. "Diffusion controlled growth of a moving sphere. The kinetics of crystal growth in a potassium perchlorate precipitation." *J. Phys. Chem.,* 65:46–9.

19 Higuchi, W. I. and Hiestand E. N. 1963. "Dissolution rates of finely divided drug powders. I. Effect of a distribution of particle sizes in a diffusion-controlled process. *J. Pharm. Sci.,* 52:67–71.

20 Pedersen, P. V. 1977. "New method for characterizing dissolution properties of drug powders." *J. Pharm Sci.,* 66(6):761–6.

21 Bates, T. R. and Sequeira, J. A. 1975. "Bioavailability of micronized griseofulvin from corn oil-in-water emulsion, aqueous suspension, and commercial tablet dosage forms in humans." *J. Pharm. Sci.,* 64(5):793–7.

22 Duncan, W. A., Macdonald, G. and Thornton, M. J. 1962. "Some factors influencing the absorption of griseofulvin from the gastro-intestinal tract." *J. Pharm. Pharmacol.,* 14:217–24.

23 Venkataram, S. and Rogers, J. A. 1984. "Characteristics of drug-phospholipid coprecipitates I: Physical properties and dissolution behavior of griseofulvin-dimyristoylphosphatidylcholine systems." *J. Pharm. Sci.,* 73(6):757–61.

24 Venkataram, S. and Rogers, J. A. 1985. "Controlled releaseof griseofulvin coprecipitates with phospholipids." *Drug Dev. Ind. Pharm.,* 11:223–38.

25 Venkataram S. 1986. "Physical characterization and bioavailability of griseofulvin: phospholipid coprecipitates." Ph.D. thesis, University of Alberta.

26 Fujii, M., Terai, H., Mori, T., Sawada, Y. and Matsumoto, M. 1988. "The properties of solid dispersions of indomethacin, ketoprofen and flurbiprofen in phosphatidylcholine." *Chem. Pharm. Bull.* (Tokyo), 36(6):2186–92.

27 Fujii, M., Harada, K., Yamanobe, K. and Matsumoto, M. 1988. "Dissolution and bioavailability of phenytoin in solid dispersion with phosphatidylcholine." *Chem. Pharm. Bull.* (Tokyo), 36(12):4908–13.

28 Nishihata, T., Chigawa, Y., Kamada, A., Sakai, K., Matsumoto, K., Shinozaki, K. and Tabata, Y. 1988. "Griseofulvin-hydrogenated soya phospholipid coprecipitates." *Drug Dev. Ind. Pharm.,* 14:1137–54.

29 Ansell, G. B. and Spanner, S. 1982. "Phosphotidylserine, phosphatidylethanolamine and phosphatidylcholine." In *Phospholipids,* J.N. Hawthorne, G.B. Ansell, Eds. Amsterdam, The Netherlands: Elsevier Biomedical Press, pp. 1–49.

30 Cevc, G. and Marsh, D. 1987. *Phospholipid Bilayers: Physical Principles and Models.* New York: John Wiley & Sons Inc., p. 441.

31 Venkataram, S. and Rogers, J. A. 1988. "Characteristics of drug-phospholipid coprecipitates. II: Bioavailability studies of griseofulvin in rats." *J. Pharm. Sci.,* 77(11):933–6.

32 Vudathala, G. K. and Rogers, J. A. 1992. "Dissolution of fludrocortisone from phospholipid coprecipitates." *J. Pharm. Sci.,* 81(3):282–6.

33 Haleblian, J. K. 1975. "Characterization of habits and crystalline modification of solids and their pharmaceutical applications." *J. Pharm. Sci.,* 64(8):1269–88.

34 Shefter, E. and Higuchi, T. 1963. "Dissolution behavior of crystalline solvated and nonsolvated forms of some pharmaceuticals." *J. Pharm. Sci.,* 52:781–91.

35 *The Merck Index,* 1976. Rahway, N. J.: Merck & Co., Inc., 9th edition, p. 534.

36 Potti, G. K., Gallelli, J. F., Akogyeram, C. O., Ahmadi, B. and Habib, M. J. 1993. "Improved dissolution characteristics of indomethacin in coprecipitates with phospholipids-II." *Drug Dev. Ind. Pharm.,* 19 (10):1221–9.

37 Biswas, M., Akogyeram, C. O., Scott, K. R., Potti, G. K., Gallelli, J. F. and Habib, M. J. 1993. "Development of carbamazepine: Phospholipid solid dispersion formulations." *J. Controlled Release,* 23:239–45.

38 Cheng, K. C., Shefter, E. and Srikrishnan, T. 1979. "Crystal structure analysis of the desolvation of the chloroform solvate of griseofulvin." *Int. J. Pharm.,* 2:81–9.

39 Vudathala, G. K. and Rogers, J. A. 1992. "Microencapsulation of solid dispersions: release of griseofulvin from griseofulvin:phospholipid coprecipitates in microspheres." *Pharm Res.,* 9(6):759–63.

40 Chiou, W. L. and Riegelman, S. 1971. "Pharmaceutical applications of solid dispersion systems." *J. Pharm. Sci.,* 60(9):1281–302.

41 Ford, J. L. 1986. "The current status of solid dispersions." *Pharm. Acta. Helv.,* 61(3):69–88.

42 Alonso, M. J., Maincent, P., Garcia-Arias, T. and Vila-Jato, J. L. 1988. "A comparative biopharmaceutical study of fresh and ageing tolbutamide-polyethyleneglycols solid dispersions." *Int. J. Pharm.,* 42:27–33.

43 Fromming, K. H. and Hosemann, R. 1985. "Stability problems under special consideration of soliddispersion of drugs." *STP Pharma,* 1:660.

44 Lach, J. L. and Chin, T. F. 1964. "Schardinger dextrin interaction IV. Inhibition of hydrolysis by means of molecular complex formation. *J. Pharm. Sci.,* 53:924–7.

45 Uekama, K., Fujinaga, T., Hirayama, F., Otagiri, M., Yamaski, M. and Seo, H. 1983. "Improvement of the oral bioavailability of digitalis glycosides by cyclodextrin complexation." *J. Pharm. Sci.,* 72(11):1338–41.

46 Kedzierewicz, F., Hoffman M. and Maincent, P. 1990. "Comparison of tolbutamide cyclodextrin inclusion compounds and solid dispersions. Physicochemical characteristics and dissolution studies." *Int. J. Pharm.,* 58:221–7.

47 Ikura H., Takeuchi, K. and Nakabachi, S. 1976. "Stabilization of active vitamin D3 derivatives." *Japan. Kokai,* 76:128–417.

48 Vudathala, G. K. and Rogers, J. A. 1991. "Effect of cholesterol on the aging of griseofulvin-phospholipid coprecipitates." *Int. J. Pharm.,* 69:13–9.

CHAPTER 4

Use of Polymers in Solid Dispersion Technology

GODFRIED OWUSU-ABABIO[1]

INTRODUCTION

THE pharmaceutical application of solid dispersion and solid solution of drugs in water-soluble carriers to improve their dissolution rates and ultimately bioavailability was first reported by
Sekiguchi and Obi in 1961 [1]. Since then, the use of solid dispersions and solid solutions in pharmaceutical dosage forms has been
studied extensively. Various carriers, notably water-soluble polymers
have been used, including polyethylene glycol (PEG) and polyvinylpyrrolidone. Further, various molecular weight grades of these polymers have been used for the preparation of solid dispersions and
solid solutions.

The technique of solid dispersion and solid solution is used to influence drug absorption and its subsequent bioavailability through
increased dissolution rate and sometimes increased solubility of the
drug. After a drug has been administered orally, it has to be delivered to its absorption site. The drug then has to dissolve before it
can be transported through membranes of the gastrointestinal tract
and ultimately reach the systemic or general circulation. Since the
drug has to be in solution in order for an absorption to be effected,
the rate of dissolution of a drug plays an important role in the biological availability of poorly water-soluble drugs. The physicochemical properties of a drug substance play the most important role

[1]College of Pharmacy, Florida A&M University, Tallahassee, FL, U.S.A.

in its solubility and also its rate of dissolution. These factors include particle size, crystalline state and state of hydration, solvation, complexation, and other reactive additives such as acids, basses, and buffers. For a given amount of drug, the smaller the particles, the greater the surface area exposed to the dissolution medium and ultimately the greater increase in the rate of dissolution. For example, the rates of dissolution of phenacetin and phenobarbital have been shown to be greatly increased by a decrease in particle size [2]. The rate of dissolution can be further increased by using various techniques to increase the rate of solution and ultimately bioavailability of drugs. These include solvent deposition, solvate formation, lyophilization, and solid dispersion. Solid dispersion was originally used to describe the dispersion of poorly water-soluble drugs in water-soluble inert polymeric carriers. Currently, it is also used to describe the dispersion of drugs in water-insoluble polymeric carriers. This chapter discusses solid dispersion of poorly water-soluble drugs using water-soluble polymers.

DEFINITIONS

- *Solid dispersion*—A solid dispersion is a system in which the concentration of the drug is in excess of its saturation solubility at room temperature. The excess drug separates as a solid phase, which is dispersed in the vehicle in crystalline or amorphous forms.
- *Solid solution*—A solid solution is a system in which the drug remains dissolved in the vehicle at room temperature and upon aging.
- *Coevaporates*—Solid dispersion is prepared by solvent removal processes.
- *Coprecipitates*—Solid dispersion is obtained when a precipitate of the drug and carrier is obtained by treating the solution containing the drug and the carrier with another solvent.

FORMULATION OF SOLID DISPERSION

The approaches used to prepare solid dispersion may involve melting of the ingredients, dissolution of the ingredients in a suitable organic solvent, or melting and subsequent dissolution in a suitable organic solvent. The various methods are described in detail below, including their advantages and disadvantages.

MELT METHOD

Of all the methods used in the preparation of solid dispersion, melting is the most widely used. Essentially, a drug is mixed with the carrier and heated slightly above the melting point of the highest melting solid until a clear liquid is formed. The liquid is allowed to solidify and is subsequently pulverized into a powder of a particular sieve size range, usually 125–250μm.

The solidification of the melt is accomplished in various ways. When the pharmaceutical application of solid dispersion and solid solutions of drugs in water-soluble carriers to improve the dissolution rates was first reported by Sekiguchi and Obi [1], the melt method was used. Sulfathiazole-urea mixture was melted above its eutectic temperature. The dispersion was solidified on an ice bath by fast cooling and pulverized to a powder. A similar fast-cooling approach was used by Alden et al. [3] to prepare polyethylene glycol-griseofulvin mixture, Jafari et al. to prepare polythylene glycol or polyvinylpyrrolidone and urea mixtures [4], Craig and Newton to characterize polyethylene glycols of various molecular weights [5], and by Alden et al. to prepare solid dispersion of polyethylene glycol-griseofulvin-sodium dodecyl sulfate system [6]. The dispersion can also be cooled through the process of spray congealing using spray-drying equipment. The melted material is sprayed onto cold metal surfaces, which forms pellets of the dispersion. This does not require grinding and, therefore, no alteration of the crystal modification of the drug occurs [7].

The melted dispersion can be cooled at room temperature. Most solid dispersions are prepared and cooled at room temperature. For example, Craig and Newton [5] prepared solid dispersions by this approach. In addition, the dispersion can be cooled at a controlled rate. Saers et al. [8] prepared solid dispersions of polyethylene glycol and *p*-aminobenzoates at a cooling rate of 10°C/minute.

Advantages

This method is very suitable for drugs and carriers that are miscible in the molten state, making melting of the ingredients very easy to accomplish. Preparing solid dispersion by the melt method is not time consuming; hence many batches of the product can be prepared in a very short time. The method is also advantageous for compounds, which do not undergo significant thermal degradation.

Disadvantages

The main disadvantages of the melt method include thermal degradation, sublimation, and polymeric transformation [9], which can affect the physicochemical properties of the drug including its rate of dissolution. It has been shown by Chiou and Niazi [10] and Ford et al. [11] that decomposition or thermal degradation is often composition-dependent and affected by melting time and the rate of cooling [12]. In order to reduce decomposition to acceptable levels, melting may be carried out at a temperature just above the highest melting component of the dispersion, which completely melts both drug and the carrier. However, it has been demonstrated by Ford et al. [13] that it is possible to prepare dispersions at just above the eutectic temperature when the carrier level is to the excess of the eutectic composition.

The temperature at which the dispersion solidifies affects crystallization rates and may alter both the size of the crystals and the hardness of the dispersion [14]. This may result in tacky or glassy and unmanageable dispersions, which will require storage at elevated temperature to facilitate hardening. Upon comm0nition of such dispersions, crystallization may be induced resulting in modification of dissolution characteristics.

If the drug and carrier display a miscibility gap such as reported by El-Banna et al. [15], which occurs in the tolbutamide-mannitol phase diagram, there may be only moderate increase in dissolution rate. This might be due to irregular crystallization and formulation of this kind of product can be a problem.

SOLVENT METHOD

The solvent method is suitable for drugs that are thermolabile and are degraded at the temperature at which the carrier will melt. The method involves dissolution of the drug and carrier in a suitable organic solvent in which both drug and carrier are soluble, followed by the removal of the solvent to form solid dispersion. This approach of preparing solid dispersion was first reported by Tachibana and Nakamura [16], where beta-carotene and polyvinylpyrrolidone were dissolved in chloroform and film casted over mercury to recover the dispersion.

Solvent removal is accomplished by various means. The most common approach is the application of reduced pressure at a fixed temperature to evaporate the organic solvent. Temperatures of 125°C for 25 minutes [17], 115°C for one hour [17], and −5°C and reduced pressure followed by drying for 12 hours in vacuum [18] have been used. In some instances, additional drying in vacuum for several hours has been used to effect complete removal of solvent following the application of reduced pressure and a fixed temperature. Spray drying is another approach by which solvent removal can be accomplished and it is probably the fastest way of removing solvent. The technique was used to prepare acetohexamine-polyvinylpyrrolidone solid dispersion [19].

The freeze-drying technique is also employed to prepare solid dispersions by removal of aqueous solutions [20].

Advantages

The procedure is suitable for drugs that are thermolabile; reduced pressure and lower temperatures can be used to evaporate solvent. For aqueous systems, frozen temperatures can be used to evaporate the solvent, which can enhance the integrity of the drug.

Disadvantages

Finding a suitable solvent that will dissolve both the drug and the carrier is very difficult and sometimes impossible. This is because most of the carriers are hydrophilic whereas most of the drugs are hydrophobic organic substances. This may be further complicated by the fact that different polymorphic forms of the same drug may be obtained if different solvents are used. Spironolactone dispersions in polyvinylpyrollidone were evaporated from solutions of ethanol, acetonitrile, and chloroform, respectively [21]. The highest dissolution rate was provided by ethanolic dispersions, whereas the chloroform dispersion provided the lowest dissolution rate. After a suitable solvent has been found, the rate of its removal is very critical in some solid dispersions, and complete removal of the solvent is even more difficult to accomplish. Plasticization of some polymers such as polyvinylpyrrolidone has occurred with the use of some solvents [22]. This made removal of the solvent extremely difficult.

It is also important that the rate of evaporation of a solvent is controlled so as to control the particle size of the drug, which in turn will affect the rate of dissolution of the drug in the solid dispersion.

The volume of organic solvent needed to dissolve a suitable amount of drug and carrier is very large in some cases, and the recovery of solvent may be economically prohibitive. For example, Chiou and Riegelman used 500 mL of ethanol to dissolve 5 g of 10% griseofulvin-polyethylene glycol solid dispersion [17].

The largest increase in dissolution rate is associated with dispersions containing molecularly dispersed drug, which results in amorphous solid dispersions. Factors that influence the formation of amorphous dispersions include the ratio of the carrier, the physical properties of the drug molecule, and the method of preparation. Sekikawa et al. [23] observed that acetohexamide-PVP 25,000 containing 70% carrier or more were amorphous and that release of drug was very rapid whereas those containing less than 70% carrier were increasingly crystalline and yielded only low dissolution rates [20]. Using drug-PVP models, Sekikawa et al. [23] quantified the mechanism of solvent removal from solid dispersions. The removal of some of the solvent due to evaporation results in a drug reaching and exceeding its solubility. As supersaturation is attained and exceeded, PVP is high enough to inhibit crystallization, which results in a dispersion containing amorphous drug. Lower concentrations of PVA cannot inhibit crystallization, resulting in crystalline drug formation in the dispersion.

POLYMERS USED IN SOLID DISPERSION

The discussion will be restricted to water-soluble polymers that increase the rate of dissolution of drugs. Two main polymers have been used in this regard, notably, polyethylene glycol and polyvinylpyrrolidone (PVP).

POLYETHYLENE GLYCOLS

The term *polyethylene glycols* essentially refers to compounds that are obtained by reacting ethylene glycol with ethylene oxide. Their molecular weights range from 200 to 300,000. Polyethylene glycols whose molecular weights are above 300,000 are commonly termed *polyethylene oxides*. Polyethylene glycols of molecular weights between 200 to 600 are clear, viscous liquids at room temperature whereas those between 900 to 8,000 are white, waxy solids.

As the molecular weight increases, the water solubility decreases. In the preparation of solid dispersions, low molecular weight polymers of this compound are used. Polymers of molecular weights ranging from 200 to 20,000 have been used extensively. It would be expected that the effect of molecular weight of the PEG on the rate of dissolution and solubility would increase or decrease in one direction. In fact the effect of molecular weight on the rate of dissolution is influenced by several factors such as the weight fraction of PEG, the solubilizing effect of PEG, the configuration of PEG molecules within the PEG matrix, the physicochemical properties of the drug, the retardation of crystallization, solid solution formation, and eutectic position [24].

Effect of PEG Molecular Weight

The dissolution rate of pure PEG decreases with increasing molecular weight [25]. This is not unexpected since, in general, an increase in molecular weight of a polymer increases the chain length and thereby decreases its solubility in a solvent. However, when the polymer is combined with a drug to prepare a solid dispersion, one of three dissolution characteristics can be observed.

The dissolution rate of the drug in the solid dispersion can be decreased with an increase in the molecular weight of PEG. This phenomenon was observed for tolbutamide [26] and indomethacin [27]. In these cases, the rate at which the polymer dissolved dictated the rate at which the drug dissolved. Lower-molecular-weight PEGs that melt at temperatures at which dissolution studies are conducted, i.e., 37°C, probably melt in the dissolution medium prior to dissolution, further increasing the rate of dissolution.

On the other hand, in some drug-PEG solid dispersion systems, the rate of dissolution decreases with molecular weight up to a certain composition of the drug above which the trend becomes irregular. An example is glutethimide-PEG solid dispersion, where a decrease in dissolution rate with increasing molecular weight was observed up to 10% drug composition [28]. Above this composition, there was no pattern in the dissolution rate as a probable result of variable crystallization.

The dissolution rate of the drug in solid dispersion can be increased by increasing the molecular weight of PEG. For example, frusemide [29] and papaverine [30] solid dispersions showed this trend. It has been suggested that higher-molecular-weight PEG (1) form more viscous solutions, which further reduces the rate of crystallization of the drug; (2) increasingly favor the incorporation of drug as solid solutions; or (3) merely flake more readily during dissolution.

The dissolution rate of the drug in a solid dispersion does not follow any pattern with the molecular weight of PEG. This phenomenon has been demonstrated with drugs such as griseofulvin [17], chloramphenicol, and testosterone [31]. In the case of chloramphenicol-PEG solid dispersion, the dissolution rate for the solid dispersions prepared with PEG was higher than those prepared with higher-molecular-weight polymers but lower than those prepared with PEG 4000. In the case of testosterone-PEG solid dispersion, the decrease in the rate of dissolution was 1,000, 20,000, 6,000 or 4,000, respectively.

POLYVINYLPYRROLIDONE

Polyvinylpyrrolidone (PVP) has a mean molecular weight ranging from about 10,000 to 700,000. It is soluble in various solvents including water, ethanol, chloroform, and isopropyl alcohol. PVP melts at a very high temperature, above 275°C, where it becomes decomposed. PVP is therefore not suitable for preparation of solid dispersion by the melt method, but only for preparation of solid dispersion by the solvent method. Typical molecular weights of PVP that are used for preparation of solid dispersion are in the range of 10,000 to 70,000.

Effect of PVP Molecular Weight

The effect of the molecular weight of PVP on the rate of dissolution of a drug is more consistent than for PEG. In general, an increase in the molecular weight of PVP will decrease the dissolution rate of most drugs. As the molecular weight of PVP increases, the rate of dissolution of the polymer itself decreases as its viscosity and swelling increase. An increase in viscosity of the PVP solution due to an increase in molecular weight decreases diffusion of drug molecules from the surface of the viscous material into the dissolution medium. The net effect is a decrease in dissolution rate of the drug and polymer. Lower-molecular-weight PVP probably undergoes a short swelling time prior to dissolution, resulting in an increase in dissolution rate of the polymer and drug [32]. The swelling phenomenon explains why, although testosterone has a higher solubility in PVP 40,000 than in PVP 11,500, the rate of dissolution of the solid dispersion of PVP 11,500 is higher than PVP 40,000 [31].

Examples of drugs whose dissolution rates have been increased with decreasing molecular weights of PVP include sulphathiazole [33], digoxin [34], sulfathiazole [35], phenytoin [36], prednisone [37], and hydrocortisone [38].

In PEG-drug solid dispersion systems, the irregular increase in dissolution rate with increasing molecular weight is thought to be due to a decrease in the crystallization rate of the drug. A similar decrease in crystallization rate was observed for PVP-sulphisoxazole dispersions, where the inhibition of crystallization was highest for PVP 40,000 followed by 360,000 and 10,000, respectively [33]. However, this did not translate into an increase in dissolution rate due to the influence of viscosity on the rate of diffusion of the drug into the dissolution medium. The rate of dissolution in decreasing order for PVP-sulphisoxazole dispersions of 3:1 composition was 10,000 followed by 40,000 and 360,000, respectively, which is consistent with the pattern of dissolution from PVP-drug dispersion systems [33]. Self-association of drug molecules within PVP solid dispersion systems has been shown to be responsible for the molecular weight effects of decreasing rate of dissolution. Self-association of drug molecules occurs more rapidly in high-molecular-weight PVP dispersion systems than lower-molecular-weight systems as demonstrated with PVP-hydrocortisone dispersion systems [38]. Since the rate of dissolution of PVP-drug dispersion systems decreases with increasing molecular weight whereas high molecular weight polymers decrease the rate of crystallization, it is possible that higher-molecular-weight dispersions age more rapidly after preparation prior to the start of dissolution studies.

Only a few cases have been reported in the literature where inconsistent dissolution rates were observed. In a PVP-nifedipine dispersion of 3:1 composition, the dissolution rate for PVP 40,000 was the highest followed by 360,000 and 10,000, respectively [39]. Likewise a decrease in the dissolution rate for PVP-indomethacine dispersion of 1:1 composition was in the order of 40,000, 10,000 and 360,000, respectively [40]. These irregular patterns are probably related to the inhibition of crystallization by these polymers. Although

PVP pyrollidone is an effective carrier and retards crystallization of many drugs, it is not an effective carrier for drugs such as caffeine or nalidixic acid [23], probably due to its inability to effectively retard crystallization.

POLYMERS AND SURFACE ACTIVE AGENTS COMBINATIONS

Surfactants lower the interfacial tension between a drug and the dissolution medium, and thereby promote wetting of the drug. The addition of surfactants to dissolution medium enhances the solubility and dissolution of drugs, such as indomethacine and phenylbutazone [41]. It is also common to add surfactant to the dissolution media to achieve consistent results during the dissolution of drugs that are poorly soluble in water. Although the incorporation of surfactants enhances dissolution in some cases, it does not have any effect on the dissolution or dispersion rate of some drugs [42]. A suitable alternative has been the incorporation of surfactants in solid dispersion to form ternary systems.

Ternary dispersion systems have higher dissolution rates than binary dispersion systems. An example is the work of Serajuddin et al. [43], where PEG and polysorbate 80, a nonionic surfactant, were combined with drug to form solid dispersion of various composition. Figure 4.1 shows that complete dissolution of drug in water occurred in 15 minutes when the ratio of the surfactant:PEG was 3:1. In the case of 1 part of surfactant to 1 part of PEG, about 60% of the drug was dissolved in 15 minutes without appreciable dissolution for the duration of the 90-minutes study. Within 90 minutes, about 40% of the drug in a binary dispersion containing no surfactant and 20% of the free drug, respectively, were dissolved. Dissolution studies in simulated gastric fluid showed slightly higher dissolution rates than in water but the pattern of dissolution did not change [43]. In a similar study where oxidipine or griseofulvin was combined with PEG and polysorbate 20 to form ternary dispersion systems, the ternary systems provided better dissolution rates than the binary systems [41].

The effect of surfactant type on the properties of a sparingly soluble drug in solid dispersion was determined for anionic, cationic and nonionic surfactants [44]. The surfactants, sodium lauryl sulfate (sodium dodecyl sulfate; SDS), dodecyltrimethyl ammonium

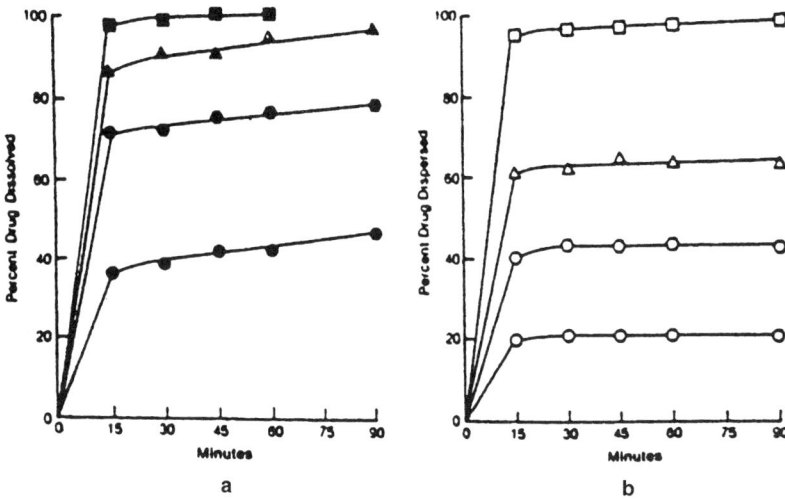

Figure 4.1 (a) Dissolution of piroxicam in simulated gastric fluid (without enzyme) from solid dispersions in PEG 1000:polysorbate 80 mixtures at 37°C. (b) Dispersion of piroxicam in water from solid dispersions in PEG 1000:polysorbate 80 mixtures at 37°C. Each capsule contained 100 mg of drug and 550 mg of vehicle. Key to ratio of PEG 1000 and polysorbate 80 in the vehicle: (●,○) 1:0 (PEG 1000 only); (⊙,⊙) 3:1; (▼,△) 1:1; and (■, □) 1:3. Each datum point represents the average of three determinations. Reprinted with permission from *J. Pharm. Sci.,* 79:464, 1990.

(DTAB), and polyethylene dodecyl ether (Brij 35), are anionic, cationic, and nonionic, respectively. When these surfactants were used to prepare ternary dispersion systems with PEG and griseoful-vin, complete instant dissolution was seen for dispersion containing 2% of SDS, and SDS was found to be the most effective surfactant. The amount of crystalline drug decreased with increasing concentration of anionic or cationic surfactants except for the nonionic surfactant.

BIOAVAILABILITY STUDIES OF POLYMER-DRUG SOLID DISPERSION SYSTEMS

PEG-DRUG SOLID DISPERSION SYSTEMS

Bioavailability studies of PEG-griseofulvin solid dispersion prepared by the melt and solvent methods have been carried out in detail both in animals and in man. For example, urinary excretion data in dogs for PEG 6000-griseofulvin dispersion containing 10% griseofulvin was 88% compared with 100, 45, and 33% for PEG 400 solution, a commercial capsule and tablet, respectively [45]. The absorption after oral administration in humans of PEG-griseofulvin containing 10% of griseofulvin prepared by the melt or solvent method was very rapid compared with absorption from micronized griseofulvin, which was irregular and incomplete [46].

Sheen et al. [47] demonstrated the effect of formulating a drug as a solid dispersion on the bioavailability of the drug in humans. The solid dispersion consisting of alpha-pentyl-3-(2-quinolinyl-methoxy)benzenemethanol and polyethylene glycol-gelucire mixture in a capsule and a tablet containing the drug was administered to human volunteers in a crossover fashion. Figure 4.2 shows that the peak serum concentration for the capsule was more than three times higher in the fasting subjects. Food intake diminished the peak concentration for the capsule, but was higher than that of the tablet.

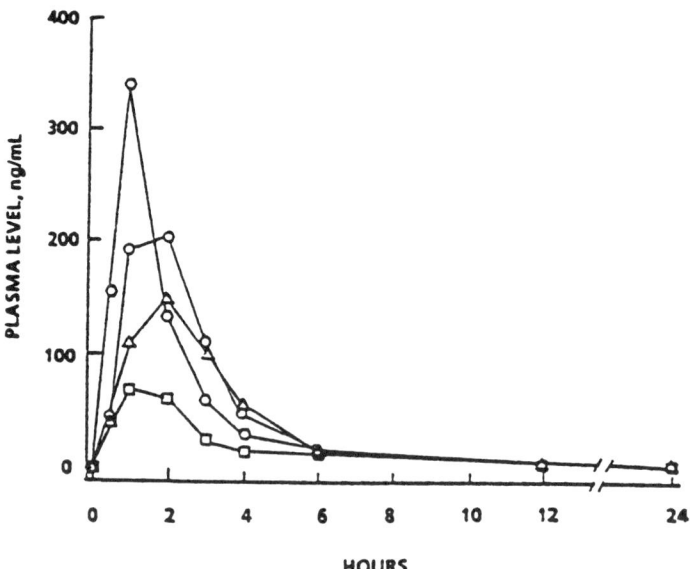

Figure 4.2 Plasma levels of alpha-pentyl-3-(2-quinolinylmethoxy)benzenemethanol in humans following oral administration of the solid dispersion (filled into soft gelatin capsule) and conventional tablet under fasted and fed regimens. Each curve represents the average plasma concentrations from eight subjects. Key: (⊙) solid dispersion after fasting: (○) solid dispersion after food; (□) tablet after fasting; (△) tablet after food. Reprinted with permission from *J. Pharm. Sci.*, 80:713, 1991.

The area under concentration time curve (AUC) without and with food for the capsule containing solid dispersion was 759 and 713 ngh/mL, respectively. This was higher than the values for the tablets without and with food of 313 and 594 ngh/mL, respectively. Vila-Jato et al. [48] reported an increase in the total amount of acetaminophen with increasing molecular weight of PEG in healthy volunteers, but the rate of absorption of the drug was unaffected.

Animal studies have revealed a similar pattern of an increase in bioavailability of solid dispersion over the drug itself. In a study with rabbits [49], a significant increase in bioavailability was observed. Table 4.1 shows that the maximum concentration and AUC for the PEG-piroxacin dispersion were 3.5 μg/mL and 217 μgh/mL, respectively, whereas 2.4 μg/mL and 168 μgh/mL were the maximum concentration and AUC, respectively, for the pure piroxacin. Similarly, mefenamic acid solid dispersion administered to animals has been found to have a significant increase in bioavailability with slight reduction in ulcerogenic activity [49]. In some cases, the rate of dissolution of the drug in the PEG solid dispersion was shown to be faster than the free drug as provided by the rapid onset of action of the drug; however, the bioavailabilities of the solid dispersions were not significantly different from the free drug. PEG-glybornuride dispersion was shown in rabbits to dissolve at a faster rate and to induce a faster decrease in glucose levels than a physical mixture with PEG [50]. The AUC for the solid dispersion was not statistically different from that of the free drug.

TABLE 4.1. Pharmacokinetic Parameters of Piroxicam Solid Dispersion After Oral Administration in Rabbits.

Formulation	No. of Rabbits	C_{max}[b] (μg/mL)	t_{max}[b] (hrs)	AUC$_{0-\infty}$ (μg.hr/mL)	K_{el}	$t_{1/2}$
Piroxicam (free drug)	6	2.4 ± 0.10	5.0	167.71 ± 8.33	0.033 ± 0.002	21.0 ± 0.85
1:2 solid dispersion of drug: PEG 4000	6	3.5 ± 0.15*	4.0	216.71 ± 7.12*	0.04 ± 0.004	17.32 ± 1.02

[a]Mean values (±SD) are shown. A significant difference from pure piroxicam, as determined by student t test is indicated by asterisk (*), $P < 0.01$.
[b]Observed peak concentration at time.
Reproduced with permission from *Drug Dev. Ind. Pharm.*, 19:739, 1993.[a]

PVP-DRUG SOLID DISPERSION SYSTEMS

In vivo evaluation of a number of PVP-drug solid dispersion has been conducted. Reserpine is one of the drugs whose PVP-solid dispersion has been shown to be more effective than the drug administered alone. PVP 40,000-reserpine solid dispersion of 5:1 composition gave a higher excretion of the drug compared to the physical mixture or the free drug, prompting a proposition in the reduction of the dose to one-third for the dispersion [51]. However, the rate and extent of absorption of reserpine from PVP-reserpine dispersion of 2:1 and 10:1 composition were not statistically different [52]. This was attributed to precipitation of excess reserpine from the aqueous fluid in the gastrointestinal tract.

PVP-tolbutamide solid dispersions have been shown to give higher blood levels than the free drug following oral administration [53]. For example, PVP-nabilone dispersion was completely absorbed from the gastrointestinal tract in humans whereas encapsulated crystalline exhibited only 60% of the bioavailability of the dispersion formulation. The dispersion formulation has been released in the United Kingdom by the Ely Lilly Company. Similarly, *in vivo* studies of PEG-rifampicin dispersion administered to human subjects showed much improved bioavailability compared to marketed products of the same drug.

Animal studies have revealed a similar increase in bioavailability after oral administration of solid dispersion. Kassem et al. [54] reported an increase in maximum concentration and AUC after oral administration of PVP-sulfisoxazole dispersion to rabbits. The maximum concentration and AUC for the solid dispersion were 103 μg/mL and 114 μgh/mL, respectively, whereas those of the pure drug were 83 μg/mL and 94 μgh/mL, respectively.

Another advantage of formulating a drug as a solid dispersion in addition to increased bioavailability is the reduction in ulcerogenic adverse effects associated with nonsteroidal antiinflammatory agents. Ramadan et al. [55] have shown that PVP significantly decreases ulcerogenic activity of mefenamic acid in animals while significantly increasing the bioavailability of the drug in solid dispersion. PVP also significantly reduced the ulcerogenic activity of mefenamic compared to PEG [55].

Just as is the case of PEG, the rate of dissolution of the drug in the PVP solid dispersion can be faster than the free drug, resulting in a rapid onset of action without any appreciable difference in bioavailability from that of the free drug. PVP-frusemide solid dispersion, being amorphous, was shown to have a significantly reduced time for maximum effect concurring with *in vitro* dissolution prediction [56]. The bioavailability of the solid dispersion and pure drug were identical.

CHARACTERIZATION OF POLYMER AND SOLID DISPERSION

To fully understand the increase in dissolution rates of solid dispersions, the application of analytical tools may be essential. Analytical tools, such as differential scanning calorimetry (DSC), scanning electron microscopy, x-ray diffraction (XD), nuclear magnetic resonance (NMR), and electron micrograph (EM), have been used to characterize PEG and its solid dispersions [3,5,6,57].

PEG in the molecular weight range used for solid dispersions are semi-crystalline, containing both ordered and amorphous regions. An increase in the molecular weight of PEG increases the melting point of the polymer and the heat of fusion. Uniformity in melting is obtained if particles of a narrow size range are used, as shown by DSC endotherms obtained by Craig and Newton [5]. This was attributed to poor thermal conductivity through the solid. The rate of cooling of melted PEG has an effect on the melting point and heat of fusion of the PEG. In studies conducted by Craig and Newton [5], the melting point of the untreated PEG was the highest followed by PEG, which had been subjected to slow cooling and fast cooling, respectively, with the corresponding molar heat of fusion decreasing in the same fashion (Tables 2 and 3 in Reference [5]). The molar heat of fusion indicates that the fast cooling results in the formation of lower ordered material leading to a decrease in crystallinity. This was confirmed by ^{13}C NMR studies, which showed a narrow peak for the fast cooled, melt compared to the slow cooled melt (Figure 4.3). The shape of the peak is thought to be related to the degree of amorphous material within the sample, and peak broadening has been ascribed to the relaxation of hydrogen atoms on the methylene groups within the ordered regions of the material. DSC and NMR can provide useful information about the polymer, which ultimately has an effect on the dissolution rate of the drug incorporated in it.

The work done by Craig and Newton explains why the fast cooling technique of solid dispersion preparation results in a less crys-

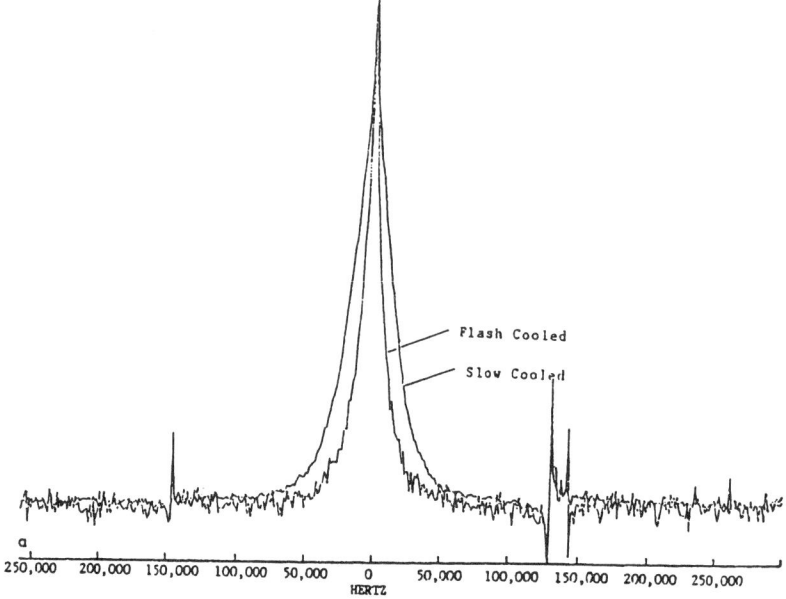

Figure 4.3 ^{1}H wideline solid state NMR studies on slow- and flash-cooled PEG samples. (a) PEG 10,000; (b) PEG 20,000. Reprinted with permission from *Int. J. Pharm.*, 74:38, 1991.

talline material, which ultimately results in an increased dissolution rate. DSC can be used to study solid dispersion to determine whether the phase is monotectic or eutectic in nature. A monotectic system is one where little or no interaction occurs whereas a eutectic system is one where significant interaction occurs. In a study where PEG mixed with the drug was melted below the melting point of the drug and slowly cooled, the DSC thermogram for the solid dispersion was the same as that of the physical mixture [57]. The phase diagrams obtained from the melting point data of the dispersions and mixtures showed no evidence of the presence of eutectics or solid solutions. EM studies showed that the drug particles remained essentially intact after fusion but were partially covered by recrystallized PEG, confirming DSC studies. In another study, PEG and griseofulvin were both melted and fast cooled [6]. X-ray diffraction studies (Figure 4 in Reference [6]) showed that the degree of crystallinity of the griseofulvin was reduced by the PEG. The NMR studies (Figure 5 in Reference [6]) essentially showed that the spectrum is a superposition of the spectra from PEG and griseofulvin. This indicates that there was no interaction between the polymer and the griseofulvin and, as a result, no solid solution or eutectic was formed. Solid solution or eutectic formation may be followed by the application of x-ray diffraction and NMR. The clearly enhanced dissolution rate of a drug in a ternary dispersion introduced by the addition of a surfactant could be explained by the formation of solid solutions. Knowledge of the interactions on a molecular level between a polymer, drug, and surfactant is of interest for the prediction of the physical and chemical properties of the ternary dispersion system. NMR is the major analytical tool of choice in acquiring knowledge of such interactions.

To use these analytical tools to explain solid solution formation, the work of Alden et al. [3] is a good example. Ternary solid dispersion system of PEG and griseofulvin containing SDS (anionic), DTAB (cationic), and Brij 35 (nonionic), respectively, were studied. They reported that x-ray powder diffractograms (Figure 6 in Reference [3]) showed a solid solution, which was similar to the finding of Alden et al. [6]. In the case of the cationic and nonionic surfactants, no solid solutions were formed although a limited solid solubility of griseofulvin in the PEG-cationic dispersion was created. The observed phenomenon is in agreement with the dissolution rate data from Sjokvist et al. [44], where increasing dissolution rates were in the order of nonionic, cationic, and anionic, respectively. Solid solution formation, therefore, enhances increased rate of dissolution.

PROBLEMS ASSOCIATED WITH PROCESSING AND STABILITY OF SOLID DISPERSION SYSTEMS

Destabilization of solid dispersion systems results in decreased dissolution rate, due to a number of factors. For example, solid dispersion systems may be destabilized through physical treatment such as pulverization and aging. Upon melting a drug and PEG, unpulverizable, sticky, metastable or glassy solids may form after cooling. Pulverization has resulted in the conversion of amorphous drugs in solid dispersion to crystalline forms. Melts containing 5–50% griseofulvin in PEG dispersion immediately after preparation were shown to be amorphous. When the melt was allowed to cool and pulverized, crystallinity was induced in 10–50% griseofulvin dispersion in PEG [3]. Serajuddin et al. [58] avoided the problem of pulverization by encapsulating the solid dispersion directly in hard gelatin capsules by the melt-filling technique. Although crystallization did not occur in one month in the dispersion with the lowest amount of drug, it did occur in the others with the increasing amount of drug.

Aging is the major problem associated with the stability of solid dispersions. Dissolution rates have been shown to decrease with age for a number of solid dispersion formulations [59–61]. As the dispersion ages, crystallite and crystallite sizes increase leading to an increase in the amount of crystalline drug. An increase in the amount of crystalline drug in turn results in a decrease in dissolution rate with age.

Although pulverization and aging increase crystallinity in solid dispersion leading to decreased dissolution rate, the problem may be overcome by the formation of solid solution. Sodium lauryl sulfate, an anionic surfactant, is known to form a complete solid solution at a certain concentration. Sjokvist et al. [44] demonstrated that solid solutions of griseofulvin in PEG containing at least 1% sodium lauryl sulfate were stable and that dissolution rates were unchanged after 12 months of storage.

REFERENCES

1 Sekiguchi, K. and Obi, N. 1961. "Studies on absorption of eutectic mixtures. I. A comparison of the behavior of eutectic mixture of sulfathiazole and that of ordinary sulfathiazole in man." *Chem. Pharm. Bull.,* 9:866–72.

2 Finholt, P. 1974. "Influence of formulation on dissolution rate." In *Dissolution Technology,* L. J. Leeson, J. T. Carstensen, Eds. Washington, D.C.: AphA, pp. 108–46.

3 Alden, M., Tegenfeldt, J. and Saers, E. S. 1993. "Structures formed by interactions in solid dispersions of the system polyethylene glycol-griseofulvin with charged and non charged surfactants added." *Int. J. Pharm.,* 94:31–8.

4 Jafari, M. R., Danti, A. G. and Ahmed, I. 1988. "Comparison of polyethylene glycol, polyvinylpyrrolidone and urea as excipients for solid dispersion systems of miconazole nitrate. *Int. J. Pharm.,* 48:207–15.

5 Craig, D. Q. M. and Newton, J. M. 1991. "Characterisation of polyethylene glycols using differential scanning calorimetry." *Int. J. Pharm.,* 74:33–41.

6 Alden, M., Tegenfeldt, J. and Sjokvist, E. 1992. "Structure of solid dispersions in the system polyethylene glycol-griseofulvin with additions of sodium dodecyl sulfate." *Int. J. Pharm.,* 83:47–52.

7 Froemming, K. H., Simons, B., Haase, J. and Hosemann R. 1978. "Problems of preparation, processing and storage of melt-embedded polyethylene glycol products." *Pharm. Ind.,* 40:967–70.

8 Saers, E. S., Nystrom, C. and Alden, M. 1993. "Psycochemical aspects of drug release. XVI. The effect of storage on drug dissolution from solid dispersions and the influence of cooling rate and incorporation os surfactant." *Int. J. Pharm.,* 90:105–18.

9 Goldberg, A. H., Gibaldi, M. and Kanig, J. L. 1965. "Increasing dissolution rates and gastrointestinal absorption of drugs via solid solutions and eutectic mixtures. I. Theoretical considerations and discussion of the literature." *J. Pharm. Sci.,* 54:1145–8.

10 Chiou, W. L. and Niazi, S. 1971. "Phase diagram and dissolution-rate studies on sulfathiazole-urea solid dispersions." *J. Pharm. Sci.,* 60:1333–7.

11 Ford, J. L., Stewart, A. F. and Rubinstein, M. H. 1979. "The assay and stability of chlorpropamide in solid dispersion with urea. *J. Pharm. Pharmac.,* 31:726–9.

12 Carcano, E. C. and Gana, I. M. 1974. "Eutectic mixtures and solid solutions of acetylsalicylic acid and urea. Stability of acetylsalicylic acid." *An. R. Acad. Farm.,* 40:487–93.

13 Ford, J. L. and Rubinstein, M. H. 1981. "Preparation, properties and aging of tables prepared from the chlorpropamide-urea solid dispersion." *Int. J. Pharm.,* 8:311–22.

14 Daabis, N. A., Abd-Elfattuh, S. and El-Banna, H. M. 1974. "Physicochemical study of drug binary systems. Part 2: Khellin-urea system." *Pharmazie,* 29: 400–4.

15 El-Banna, H. M., Daabis, N. A., Mortada, L. M. and Abd-Elifattuh, S. 1975. "Physicochemical study of drug binary systems. Part 3: Tolbutamide-urea and tolbutamide-mannitol systems." *Pharmazie,* 30:788–91.

16 Tachibana, T. and Nakamura, N. 1965. "A method of preparing an aqueous colloidal dispersion of organic materials by using water-soluble polymers: dispersion of β-carotene by poly (vinyl pyrrolidinone)." *Kolloid-Z.,* 203:130–3.

17 Chiou, W. L. and Riegelman, S. 1969. "Preparation and dissolution characteristics of several fast release solid dispersions of griseofulvin." *J. Pharm. Sci.,* 58:1505–10.

18 Malone, M. H., Hochman, H. L. and Nieforth, K. A. 1966. "Desoxycholic acid enhancement of orally administered reserpine." *J. Pharm. Sci.,* 55:972–4.

19 Graf, E., Beyer, C. and Abdallah, O. 1982. "Solid dispersions of acetohexamide. 3. Acetohexamide-poly (vinyl pyrrolidinone). Methods of preparation and solvents used." *Acta Pharm. Technol.,* 28:225–30.

20 Sekikawa, H., Fakuda, W., Takada, M., Ohtani, K., Arita T. and Nakuno, M. 1979. "Dissolution behaviors and gastrointestinal absorption of tolbutamide in tolbutamide-polyvinylpyrrolidone coprecipitate." *Chem. Pharm. Bull.,* 27:31–7.

21 El-Dalsh, S. S., El-Sayed, A. A., Badawi, A. A. and Fouli, A. 1982. "Effect of the solvent used in coprecipitation on the dissolution rate of spironolactone." *Pharmazie,* 37:606–7.

22 Stevens, L. A. and Padfield, J. M. 1977. "Co-precipitated systems of salicylic acid (SSA) with polyvinyl pyrrolidone (PVP): physical and bioavailability studies." *Congr. Int. Technol. Pharm.,* 5:135–41.

23 Sekikawa, H., Nakano, M. and Arita, T. 1978. "Inhibitory effect of polyvinylpyrrolidone on the crystallization of drugs." *Chem. Pharm. Bull.,* 26:118–26.

24 Ford, J. L. 1980. *Physical, Dissolution and Formulation Properties of Solid Dispersions.* Liverpool, England: Liverpool Polytechnic, pp. 1–10.

25 Corrigan, O. J., Murphy, C. A. and Timoney, R. F. 1979. "Dissolution properties of polyethylene glycols and polyethylene glycol-drug systems." *Int. J. Pharm.*, 4:67–74.

26 Miralles, M. J., McGinty, J. W. and Martin, A. 1982. "Combined water-soluble carriers for coprecipitates of tolbutamide." *J. Pharm. Sci.*, 71:302–4.

27 Krasowska, H., Krowczynski, L. and Kocelak, E. 1975. "Influence of poly(oxyethylene) glycol molecular weight on the rate of release of indomethacin from suppositories." *Farm. Pol.*, 31:291–5.

28 Ford, J. L. 1984. "The influence of polyethylene glycol molecular weight variation on the properties of glutethimide-polyethylene glycol solid dispersions." *Pharm. Acta. Helv.*, 59:280–8.

29 Shin, S. 1979. "Studies on hydrophobic drug-soluble carrier coprecipitates. I. Dissolution characteristics of furosemide-polymer coprecipitates." *Arch. Pharm. Res.*, 2:35–47.

30 El-Gindy, N. A., Karara, A. H. and El-Khalck, M. M. A. 1976. "Enhanced dissolution rate of papaverine via the solid dispersion technique." *Sci. Pharm.*, 44:283–8.

31 Hoelgaard, A. and Moeller, N. 1975. "Dissolution characteristics of drugs from solid dispersions." *Arch. Pharm. Chemi. Sci. Ed.*, 3:65–72.

32 Nogami, H., Nagai, T. and Kondo, A. 1970. "Dissolution kinetics of polyvinylpyrrolidone." *Chem. Pharm. Bull.*, 18:1185–90.

33 Sekikawa, H., Nakano, M. and Arita, T. 1978. "Dissolution behavior and gastrointestinal absorption of sulfisoxzole in a sulfisoxazole-polyvinylpyrrolidone coprecipitate." *Yakugaku Z.*, 98:62–6.

34 Ammar, H. O., Kassem, M. A., Salana, H. A. and El-Ridy, M. S. 1980. "On the dissolution of digoxin." *Pharm. Ind.*, 42:757–61.

35 Simonelli, A. P., Mehta, S. C. and Higuchi, W. I. 1969. "Dissolution rates of high energy polyvinylpyrrolidone (PVP)-sulfathiazole coprecipitates." *J. Pharm. Sci.*, 58:538–49.

36 Sekikawa, H., Fujiwara, J., Naganuma, T., Nakano, M. and Arita, T. 1978. "Dissolution behavior and gastrointestinal absorption of phenytoin in phenytoin-polyvinylpyrrolidone coprecipitates." *Chem. Pharm. Bull.*, 26:3033–9.

37 Merkle, H. P. 1981. "Studies on the dissolution process in drug/poly (vinylpyrrolidone) coprecipitates." *Acta Pharm. Tech.*, 27:193–203.

38 Merkle, H. P. 1982. "Aging of coprecipitates: self association as observed by dynamic dialysis of supersaturated aqueous solution of hydrocortisone." *Pharm. Acta Helv.*, 57:160–3.

39 Sugimoto, I., Kuchiki, A., Nakagawa, H., Tohgo, K., Kondo, S., Iwane, I. and Takahashi, K. 1980. "Dissolution and absorption of nifedipine from nifedipine-polyvinylpyrrolidone coprecipitate." *Drug Dev. Ind. Pharm.*, 6:137–60.

40 Takayama, K., Namba, N. and Nagai, T. 1981. "Analysis of interfacial transfer of indomethacin following dissolution of indomethacin/poly vinylpyrrolidone coprecipitates." *Chem. Pharm. Bull.*, 29:2718–21.

41 Tripathi, M., Kohoil, D. V. and Uppadhyay, R. K. 1992. "Enhancement of solubility and dissolution of indomethacin and phenylbutazone by cholic and deoxycholic acid conjugates." *Drug. Dev. Ind. Pharm.*, 18:135–41.

42 Veiga, M. D., Escobar, C. and Bernad, M. J. 1993. "Dissolution behaviour of drugs from binary and tertiary systems." *Int. J. Pharm.*, 93:215–20.

43 Serajuddin, A. T., Sheen, P. C. and Augustine, M. A. 1990. "Improved dissolution of a poorly water-soluble drug from solid dispersions in polyethylene glycol:polysorbate 80 mixtures." *J. Pharm. Sci.,* 79:463–4.

44 Sjokvist, E., Nystrom, C., Alden, M. and Caram-Lelham, N. 1992. "Pysicochemical aspects of drug release. XIV. The effects of some ionic and non-ionic surfactants on properties of a sparingly soluble drug in solid dispersions." *Int. J. Pharm.,* 79:123–33.

45 Chiou, W. L. and Riegelman, S. 1970. "Oral absorption of griseofulvin in dogs: Increased absorption *via* solid dispersion in polyethylene glycol 6000." *J. Pharm. Sci.,* 59:937–42.

46 Chiou, W. L. and Riegelman, S. 1971. "Absorption characteristics of solid dispersed and micronized griseofulvin in man." *J. Pharm. Sci.,* 60:1376–80.

47 Sheen, P. C., Kim, S. I., Petillo, J. J. and Serajuddin, A. T. 1991. "Bioavailability of a poorly water-soluble drug from tablet and solid dispersion in humans." *J. Pharm. Sci.,* 80:712–4.

48 Vila-Jato, J. L., Blanco, J. and Alonso, M. J. 1986. "The effect of the molecular weight of polyethylene glycol on the bioavailability of paracetamol-polyethylene glycol solid dispersions." *J. Pharm. Pharmacol.,* 38:126–8.

49 Bhattacharyya, M., Basu, S. K., Gupta, B. K., Ghosal, S. K., Mandal, S. C. and Chattaraj, S. C. 1993. "Formulation and in vivo characterization of solid dispersions of piroxicam." *Drug. Dev. Ind. Pharm.,* 19:739–47.

50 Vila-Jato, J. L. and Alonso, M. J. 1986. "Biopharmaceutical study of glybornuride-polyethylene glycol systems." *Drug Dev. Ind. Pharm.,* 12:701–11.

51 Stupak, E. L. and Bates, T. R. 1972. "Enhanced absorption and dissolution of reserpine from reserpine-polyvinylpyrrolidone coprecipitates." *J. Pharm. Sci.,* 61:400–4.

52 Stupak, E. I., Rasenburg, H. A. and Bates, T. R. 1974. "Biopharmaceutical physicochemical studies on reserpine-poly(vinylpyrrolidone) coprecipitates." *J. Pharmacokinet. Biopharm.,* 2:511–24.

53 Said, S. A., El-Fatatry, H. M. and Geneidi, A. S. 1974. "Co-precipitates of tolbutamide with poly(vinylpyrrolidone) and fusion mixtures with macrogel." *Aust. J. Pharm. Sci.,* 3:42–5.

54 Kassem, M. A., El-Ridy, M. S. and Khairy, L. M. 1987. "Pharmacokinetics of sulfisoxazole solid dispersions in rabbits." *Drug Dev. Ind Pharm.,* 13:1171–96.

55 Ramadan, E. M., Abd El-Gawad, H. and Nouh, A. T. 1987. "Bioavailability and erosive activity of solid dispersions of some non-steroidal anti-inflammatory drugs." *Pharm. Ind.,* 49:508–13.

56 Doherty, C. and York, P. 1989. "The *in-vitro* pH-dissolution dependence and *in-vivo* bioavailability of frusemide-PVP solid dispersions." *J. Pharm. Pharmacol.,* 41:73–8.

57 Craig, D. Q. and Newton, J. M. 1991. "Characterisation of polyethylene glycols using solution calorimetry." *Int. J. Pharm.,* 74:43–8.

58 Serajuddin, A. T., Sheen, P. C., Mufson, D., Bernstein, D. F. and Augustine, M. A. 1988. "Effect of vehicle amphiphilicity on the dissolution and bioavailability of a poor water-soluble drug from solid dispersions." *J. Pharm. Sci.,* 77:414–7.

59 Khalil, S. A. H. and Mortada, I. M. 1978. "Decreased dissolution of phenylbutazone-PEG 6000 solid dispersions upon storage." *J. Drug Res. Egypt,* 10:141–50.

60 Ford, J. L. and Rubinstein, M. H. 1979. "Aging of indomethacin-polyethylene glycol 6000 solid dispersion." *Pharm. Acta. Helv.,* 54:353–8.

61 Ford, J. L. and Rubinstein, M. H. 1980. "Formulation and aging of tablets prepared from indomethacin-polyethylene glycol 6000 solid dispersions." *Pharm. Acta. Helv.,* 55:1–7.

CHAPTER 5

Cyclodextrins in Solid Dispersions

SURESH VENKATARAM[1]
GURU V. BETAGERI[2]

INTRODUCTION

THE history of cyclodextrins in the pharmaceutical industry dates back to 1891 with the findings of Villiers [1] and subsequent research by Schardinger [2] in 1903, who recognized the role of the bacterium *Bacillus macerans* in the formation of crystalline compounds called *cyclodextrins,* which occur in three forms α, β, and γ. Scientific understanding and advancements in starch and glucose enzymology not only resulted in commercialization of large-scale production of cyclodextrins but also in the characterization and synthesis of newer cyclodextrins. Today, with the introduction of a number of cyclodextrins with varying physicochemical properties, their use is increasing in formulation research. Cyclodextrins are primarily used to enhance solubility, chemical protection, taste masking, and improved handling by the conversion of liquids into solids by entrapment.

STRUCTURAL FEATURES

Cyclodextrins are cyclic oligosaccharides derived from starch by the action of the enzyme cycloglycosyl transferase. Pharmaceutical-

[1]Patheon Inc., Mississauga, Ontario, Canada.
[2]Department of Pharmacal Sciences, School of Pharmacy, Auburn University, Auburn, AL, U.S.A.

relevant cyclodextrins are cyclic oligosaccharides composed of dextrose units joined through a 1–4 bond. The α, β, and γ are composed of 6, 7, and 8 glucopyranose units, respectively [3]. The usual orientation of the functional groups provides a hydrophobic interior and a hydrophilic exterior similar to micelles, where aggregation of individual surface active monomers has to occur to provide this environment. The orientation of primary and secondary hydroxyl groups on the outside and the etheroxide linkages along with hydrogen and carbon on the inside. The hydrophobic interior is frequently referred to as the *cavity* of the cyclodextrin molecule, which has the overall shape of a truncated cone. One of the two rims of the cone is formed by the primary hydroxyl groups attached to the C-6 carbons of glucopyranose, the other rim is the result of secondary hydroxyl groups attached to the C-2 and C-3 carbons of glucopyranose. While the height (7.9 A_i of the truncated cone remains constant for all three types of cyclodextrins, the size of the cavity increases with the number of glucopyranose units. Thus a-cyclodextrins have the smallest cavity, measuring 4.7 A to 5.3 A_i, with a capture volume or cavity volume of approximately 100 mL/mole. The corresponding numbers for β- and γ-cyclodextrins are 6.0 to 6.5 A_i and 150 mL/mole, and 7.5 A to 8.3 A_i and 200 mL/mole, respectively. The external dimensions of cyclodextrin molecules, which are not a very important parameter for a formulator, are 14.6 A, 15.4 A and 17.5 A for α-, β-, and γ-cyclodextrins, respectively.

METHODS OF DRUG INCORPORATION IN SOLID DISPERSIONS

In the *melting method of preparation,* the carrier is heated to a temperature just above its melting point and the drug is incorporated into the matrix. The mixture is cooled with constant stirring to homogeneously disperse the drug throughout the matrix. Several mechanisms may operate during the process of dispersion. If the drug has a high degree of solubility in the carrier, it may remain "dissolved" in the solid state, yielding what is known as a solid solution. Particle-size reduction under these conditions proceeds to the ultimate level, leading to molecular dispersion of the drug in the carrier matrix. These systems show very high drug-dissolution rates compared to control samples. If, on the other hand, the solubility of the drug in the solid state is not high, crystallites of the drug become dispersed in the matrix. Such systems show only moderate increases in dissolution rates. A third mechanism is the conversion of a drug to an amorphous form in the presence of the matrix, again exhibiting different dissolution rates and solubility. Other contributing factors that may play a role are solubilizing effect conferred by the carrier itself, improved wetting or decreased surface hydrophobicity, complexation, and crystallization of the drug in a metastable polymorphic form of altered thermodynamic properties. An important limitation of the fusion method of preparation is the exposure of drugs to elevated temperatures, particularly if the carrier is a high-melting solid and the drug is heat sensitive.

In the *solvent method of preparation,* the carriers and the active ingredients are dissolved in a suitable organic solvent and the solvent is evaporated at an elevated temperature or under vacuum. As the solvent is being removed, supersaturation occurs followed by simultaneous precipitation of the constituents resulting in a solid residue called a *coprecipitate.* The coprecipitate is then dried in a vacuum to eliminate any solvent freely adhering to the particle surface. However, there is a possibility of the formation of a solvate with any of the constituents, which could hold some solvent within the crystal lattice. This poses a problem in terms of pharmaceutical acceptance since most of the solvents used are nonaqueous (organic). Hence, removal of even trace amounts of the solvent is implied. Highly sensitive techniques such as differential scanning calorimetry (DSC), differential thermal analysis (DTA), and thermogravimetric analysis (TGA), and less sensitive procedures like gravimetry and spectroscopy can be used to demonstrate complete solvent removal.

In the *melt-solvent method,* a carrier/s is/are melted and the drug/s is/are incorporated in the form of a solution. If the carrier is capable of holding a certain proportion of liquid while maintaining its solid properties and if the liquid is innocuous, the need for removing the solvent is eliminated. Otherwise, this method faces the same criticism of solvent retention as described before. This method has been adopted by a number of researchers [4].

Although there are advantages and disadvantages associated with these methods, the choice of a method of preparation could affect the intended purpose of solid dispersion formulations. Najib and

Salem [5] have shown that ibuprofen-dispersions prepared by the fusion method gave higher solubilities than those prepared by the solvent method. Similarly, higher dissolution rates were obtained for sulfamethoxazole-PEG [6] or griseofulvin-PEG [7] dispersions prepared by the fusion method as compared to coprecipitation. However, Jafari et al. [8] have reported comparable results from dispersions prepared by coprecipitation and fusion methods.

The inclusion complex of paracetamol and β-CD has been reported with increased dissolution rates [9]. In pharmaceutical dosage forms the application of drug inclusion complexes is difficult for high-dosage drugs. Therefore, preparation of solid dispersion of such drugs using CD is more appropriate.

MECHANISM OF DRUG RELEASE

The rational design of formulations that take advantage of cyclodextrin inclusion complexation requires an understanding of the relationship between intrinsic drug solubility, the magnitude of the binding constant for the inclusion complex, and dilution effects [3]. The effect of dilution with plasma on hydroxypropyl-β-cyclodextrin (HP-β-CD) complexes of naproxen or flurbiprofen was reported by Frijlink et al. [10]. They found experimentally that only small fractions of the drugs remained bound to the cyclodextrin in plasma *in vitro*. This effect was due not only to dilution but also to the competition between the albumin binding of the two drugs and cyclodextrin binding. Also contributing to the low fraction bound was displacement of the drugs from cyclodextrin by a competing agent, plasma cholesterol.

The importance of changes in the ratio of free to complexed drug upon dilution of a sparingly water-soluble drug in a cyclodextrin complex depends on the phase-solubility behavior of the system. For example, if the cyclodextrin complex of a drug results from 1:1 interaction, there is a linear increase in drug solubility with increasing cyclodxtrin concentration. Therefore, dilution of a true solution of the drug/cyclodextrin complex will not result in drug precipitation regardless of the extent of dilution [3]. Precipitation of the drug may occur on dilution, however, if there is a nonlinear relationship between drug solubility and cyclodextrin concentration. Similar precipitation of drug could also occur *in vivo* when such solutions are injected intravenously or by other parenteral routes.

ORAL ADMINISTRATION OF CYCLODEXTRINS

The bioavailability of poorly water-soluble drugs can be increased by increasing the rate and extent of dissolution of drug. Cyclodextrins play a major role in achieving this goal. Other advantages to employing cyclodextrins are: increasing the stability of drug, release profile during gastrointestinal (GI) transit through modification of drug release site and time profile, decreasing local tissue irritation, and masking unpleasant taste. The use of cyclodextrins to increase the oral bioavailability of a drug through the increase of apparent drug dissolution rate has usually involved the use of the parent cyclodextrins or their more water-soluble derivatives. Such applications have been successful when the rate-limiting step in drug absorption is dissolution of the drug itself and not absorption across the GI mucosa.

Many of the studies do not correlate the increased bioavailability with increased dissolution rate. Instead, most reports include the administration of uncomplexed drug or a commercial formulation to demonstrate the relative bioavailability. It is important to note that few studies have investigated whether apparent cyclodextrin-dependent increases in oral bioavailability are solely a result of increasing the rate at which drug is available for absorption or whether the cyclodextrin induced changes in mucosal permeability. In contrast to to the use of hydrophilic cyclodextrins to increase oral bioavailability, the use of cyclodextrins to extend the release of drugs during GI transit usually has involved the use of more hydrophobic cyclodextrin derivatives while studies aimed at alleviating unwanted effects resulting from the low pH of the stomach have employed cyclodextrin derivatives which exhibit pH dependent solubility behavior. All these studies provide valuable insight into the potential uses of cyclodextrins in oral drug delivery [3].

It has been demonstrated that small amounts of β- and γ-CD were absorbed by intestinal segments under bile duct ligation during in situ loop perfusion studies in rats [11,12]. Therefore, the possibility of intestinal absorption of the cyclodextrins should not be totally disregarded. Others have also observed the absorption of small amounts of cyclodextrin through the GI tract of rats when given alone [13,14] or as complex with a drug [15,16]. A review of the absorption-enhancing effects of cyclodextrins given by various routes including oral administration has been reported [11]. While it is correct to assume that the largest impact of cyclodextrins on oral drug absorption results from an increased rate of drug dissolution of the complexed drug, the magnitude of the stability constant between the drug and the cyclodextrin should also be considered since only the free uncomplexed drug is absorbed through the GI mucosa. However, cyclodextrins may affect drug absorption through modification of the mucosal membrane [11]. Free cyclodextrin may remove membrane components [12,17], thereby modifying the transport properties of the membrane and facilitating absorption of the drug. Such properties would not necessarily be hindered by derivatization of the cyclodextrin [18].

The physical and chemical mechanisms by which cyclodextrins increase dissolution rate and bioavailability have been reviewed by Szetjli [19]. According to the model employed by the author, at drug-to-cyclodextrin ratios of unity an increase in peak blood levels and bioavailability over that of cyclodextrin-free formulation would be predicted over a range of reasonable stability constants. At high-stability constants and in the presence of molar excesses of cyclodextrin, however, the model suggests that the peak blood levels and bioavailability attained may be lower for a formulation containing cyclodextrin relative to one containing only drug. This effect would be exaggerated if the drug had a narrow absorption window. It should be noted that the model does not include competition for the cyclodextrin cavity by other molecules present in the GI tract. The limitation of oral dosage forms containing cylcodextrins with regard to total dosage unit mass and drug content has been described by the same author. The amount of drug that can be delivered in a unit dose is limited due to the difference in molecular weight of the drug and cyclodextrins and less than mole-per-mole binding of drug and cyclodextrin. This effect is important, especially since many of the hydrophilic cyclodextrin-derivative drug complexes cannot readily or economically be isolated as the pure complex. In many cases, therefore, a 500 mg tablet may only contain 5–25 mg of active drug, assuming that no other excipients in addition to the cyclodextrin are present [20].

The increases in the oral bioavailability of a drug from cyclodextrin containing formulation predominantly result from increases in the apparent solubility of the drug. Recent investigations have used more hydrophilic derivatives of β-CD. Increased oral bioavailability in beagle dogs of the weak base cinnarizine from formulations containing HP-β-CD and SBE4-β-CD has been reported by Jarvinen et al. [21]. Similar studies have been reported with antiinflammatory activity of piroxicam [22] and hamycin [23]. HP-β-CD was shown to double the bioavailability in rats of carbamazepine [24] and increase the absorption in dogs by about six times that of the commercial Tegretol formulation [25]. In other studies, HP-β-CD was shown to double the levels of ellagic acid, an antimutagenic and anticarcinogenic natural product analog, in the lungs of mice [26] and rabbits [27]. Physical mixture of paracetamol with β-CD and its solid dispersions showed increased solubility and dissolution rates from tablets compared to plain drug [20]. The study also indicated that the process of preparing solid dispersions with β-CD influenced the different increases of solubility, e.g., spray-dried products exhibited an increased rate compared with the kneading process. Dissolution rates of paracetamol from tablets made with the kneading process resulted in the largest increase in the rate of paracetamol dissolution.

REFERENCES

1 Villiers, A. 1891. "Sur la fermentation de la fecule par l'action du ferment butyrique." *C.R. Hebd. Seances Acad. Sci.,* 112:536–8.

2 Schardinger, F. 1903. "Ueber die Zulassigkeit des Warmhaltens von zum GenuB bestimmten Nahrungsmitteln mittelst Warme speichernder Apparate, sog. Thermophore." *Z. Wien. Klin. Wochenschr.,* 16:468–74.

3 Rajewski, R. A. and Stella, V. J. 1996. "Pharmaceutical applications of cycloductions. 2. *In vivo* drug delivery." *J. Pharm. Sci.,* 85:1142–69.

4 Fernandez, M., Rodriguez, I. C., Margarit, M. V. and Cerezo, A. 1992. "Characterization of solid dispersions of piroxicam/polyethylene glycol 4000." *Int. J. Pharm.,* 84:197–202.

5 Najib, N. M. and Salem, M. A. 1987. "Release of ibuprofen from polyethylene glycol solid dispersions: Equilibrium solubility approach." *Drug Dev. Ind. Pharm.,* 13:2263–75.

6 Singla, A. K. and Vijan, T. 1990. "Dissolution of sulfamethoxazole from polyethylene glycols and polyvinyl pyrrolidone solid dispersions." *Drug Dev. Ind. Pharm.,* 16:875–82.

7 Sjokvist, E. and Nystrom, C. 1988. "Physicochemical aspects of drug release. VI. Drug dissolution rate from solid particulate dispersions and the importance of carrier and dry particle properties." *Int. J. Pharm.,* 47:51–66.

8 Jafari, M. R., Danti, A. G. and Ahmed, I. 1988. "Comparison of polyethylene glycol, polyvinylpyrrolidone and urea as excipients for solid dispersion systems of miconazole nitrate." *Int. J. Pharm.,* 48:207–15.

9 Lin, S. and Kao, Y.-H. 1989. "Solid particulates of drug-β-cyclodextrin inclusion complexes directly prepared by a spray-drying technique." *Int. J. Pharm.,* 56:249–59.

10 Frijlink, H. W., Visser, J., Hefting, N. R., Oosting, R., Meijer, D. K. and Lerk, C. F. 1990. "The pharmakokinetics of beta-cyclodextrin and hydroxypropyl-beta-cyclodextrin in the rat." *Pharm. Res.,* 7:1248–52.

11 Uekama, K., Hirayama, F. and Irie, T. 1994. "Application of cyclodextrins in pharmaceutical preparations." *Drug Targeting Delivery,* 3:411–56.

12 Irie, T., Tsunenari, Y., Uekama, K. and Pitha, J. 1988. "Effect of bile on the intestinal absorption of α-cyclodextrin in rats." *Int. J. Pharm.,* 43:41–4.

13 Gerloczy, A., Fonagy, A., Keresztes, P., Perlaky, L. and Szejtli, J. 1985. "Absorption, distribution, excretion and metabolism of orally administered ^{14}C-β-cyclodextrin in rats." *Arzneim.-Forsch.,* 37:1042–7.

14 Poelma, F. G. J., Tukker, J. J., Hilbers, H. W. and Jansen, A. C. 1989. "I. Intestinal absorption of drugs. II. The effect of inclusion in cyclodextrins on the absorption of dantrolene." *J. Inclusion Phenom. Mol. Recognit. Chem.,* 7:423–30.

15 Koizumi, K., Kubota, Y., Okada, Y. and Utamura, T. 1985. "Microanalysis of β-cyclodextrin in plasma by high-performance liquid chromatography." *J. Chromatogr.,* 341:31–41.

16 Nakanishi, K., Masada, M., Nadai, T. and Miyajima, K. 1989. "Effect of the interaction of drug-β-cyclodextrin complex with bile salts on the drug absorption from rat small intestinal lumen." *Chem. Pharm. Bull.,* 37:211–4.

17 Nakanishi, K., Nadai, T., Masada, M. and Miyajima, K. 1992. "Effect of cyclodextrins on biological membrane. II. Mechanism of enhancement on the intestinal absorption of non-absorbable drug by cyclodextrins." *Chem. Pharm. Bull.,* 40:1252–6.

18 Bodor, N. S., Huang, M. and Watts, J. D. 1995. "Theoretical studies on the structures of natural and alkylated cyclodextrins." *J. Pharm. Sci.,* 84:330–6.

19 Szetjli, J. 1988. "Cyclodextrins in pharmaceuticals." In *Cyclodextrin Technology,* J. E. D. Davies, Ed.: Dordrecht, The Netherlands: Kluwer Acad. Publ. pp. 186–306.

20 Tasic, L. J. M., Jovanovic, M. D. and Djuric, Z. R. 1992. "The influence of β-cyclodextrin on the solubility and dissolution rate of paracetamol solid dispersions." *J. Pharm. Pharmacol.,* 44:52–5.

21 Jarvinen, T., Jarvinen, K., Schwarting, N. and Stella, V. J. 1995. "β-cyclodextrin derivatives, SBE4-β-Cd and HP-β-CD, increase the oral bioavailability of cinnarizine in beagle dogs." *J. Pharm. Sci.,* 84:295–9.

22 Reddy, D. N., Singh, U. V., Pandey, S. and Udupa, N. 1992. "Evaluation of piroxicam oral and transdermal preparations." *Indian J. Pharm. Sci.,* 54:152–4.

23 Singh, U. V., Pandey, S. and Udupa, N. 1993. "Effect of β-cyclodextrin on therapeutic activity of hamycin." *Indian J. Hosp. Pharm.,* 30:165–6.

24 Choudhary, S. and Nelson, K. F. 1992. "Improvement of oral bioavailability of carbamazepine by inclusion in 2-hydromypropyl-β-cyclodextrin." *Int. J. Pharm.,* 85:175–80.

25 Betlach, C. J., Gonzalez, M. A., McKiernan, B. C., Neff-Davis, C. and Bodor, N. 1993. "Oral pharmacokinetics of carbamazepine in dogs from commercial tablets and a cyclodextrin complex." *J. Pharm. Sci.,* 82:1058–60.

26 Hostetler, J. S., Hanson, L. H. and Stevens, D. A. 1993. "Effect of hydroxypropyl-β-cyclodextrin on efficacy of oral itraconzole in disseminated murine cryptococcosis." *J. Antimicrob. Chemother.,* 32:459–63.

27 Patterson, T. F., Fothergill, A. W. and Rinaldi, M. G. 1993. "Efficacy of itraconazole solution in a rabbit model of invasive aspergillosis." *Antimicrob. Agents Chemother.,* 37:2307–10.

Index

BASIC AND CLINICAL DERMATOLOGY

Series Editors

ALAN R. SHALITA, M.D.
Distinguished Teaching Professor and Chairman
Department of Dermatology
SUNY Downstate Medical Center
Brooklyn, New York

DAVID A. NORRIS, M.D.
Director of Research
Professor of Dermatology
The University of Colorado
Health Sciences Center
Denver, Colorado